FM 20-15

US ARMY FIELD MANUAL

TENTS

AND

TENT PITCHING

1956

CIVILIAN REFERENCE EDITION

UNABRIDGED GUIDEBOOK TO INDIVIDUAL AND LARGE MILITARY-STYLE WALL SHELTERS,
TEMPORARY STRUCTURES, AND CANVAS CARE

U.S. DEPARTMENT OF THE ARMY

Doublebit Press

New content, introduction, cover design, and annotations
Copyright © 2020 by Doublebit Press. All rights reserved.

Doublebit Press is an imprint of Eagle Nest Press
www.doublebitpress.com
Cherry, IL, USA

Original content under the public domain; unrestricted for civilian distribution.
Originally published in 1956 by the U.S. Department of the Army.

This title, along with other Doublebit Press books are available at a volume discount for youth groups, clubs, or reading groups. Contact Doublebit Press at info@doublebitpress.com for more information.

Military Outdoors Skills Series: Volume 10

Doublebit Press Civilian Reference Edition ISBNs
Hardcover: 978-1-64389-162-0
Paperback: 978-1-64389-163-7

Doublebit Press, or its employees, authors, and other affiliates, assume no liability for any actions performed by readers or any damages that might be related to information contained in this book. Some of the material in this book may be outdated by modern standards. This text has been published for historical study and for personal literary enrichment. Remember to be safe with any activity that you do in the outdoors and to help do your part to preserve and be a good steward of our great American wild lands.

The Military Outdoors Skills Series
Historic Field Manuals and Military Guides
on Outdoors Skills and Travel

Military manuals contain essential knowledge about outdoors life, thriving while in the field, and self-sufficiency. Unfortunately, many great military books, field manuals, and technical guides over the years have become less available and harder to find. These have either been rescinded by the armed forces or are otherwise out of print due to their age. This does not mean that these manuals are worthless or "out of date" – in fact, the opposite is true! It is true that the US Military frequently updates its manuals as its protocols frequently change based on the current times and combat situations that our armed services face. However, the knowledge about the outdoors over the entire history of military publications is timeless!

By publishing the **Military Outdoors Skills Series**, it is our goal at Doublebit Press to do what we can to preserve and share valuable military works that hold timeless knowledge about outdoors life, navigation, and survival. These books include official unrestricted texts such as army field manuals (the FM series), technical manuals (the TM series), and other military books from the Air Force, Navy, and texts from before 1900. Through remastered reprint editions of military handbooks and field manuals, outdoors enthusiasts, bushcrafters, hunters, scouts, campers, survivalists, nature lore experts, and military historians can preserve the time-tested skills and institutional knowledge that was learned through hard lessons and training by the U.S. Military and our expert soldiers.

Soldiers were the original campers and survivalists! Because of this, military field manuals about outdoors life contain essential knowledge about thriving in the wilds. This book is not just for soldiers!

This book is an important contribution to outdoors literature and has important historical and collector value toward preserving the American outdoors tradition. The knowledge it holds is an invaluable reference for practicing skills related to thriving in the outdoors. Its chapters thoroughly discuss some of the essential building blocks of outdoors knowledge that are fundamental but may have been forgotten as equipment gets fancier and technology gets smarter. In short, this book was chosen for Historic Edition printing because much of the basic skills and knowledge it contains could be forgotten or put to the wayside in trade for more modern conveniences and methods.

Although the editors at Doublebit Press are thrilled to have comfortable experiences in the woods and love our high-tech and light-weight equipment, we are also realizing that the basic skills taught by the old experts are more essential than ever as our culture becomes more and more hooked on digital technology. We don't want to risk forgetting the important steps, skills, or building blocks involved with thriving in the outdoors. This Civilian Reference Edition reprint represents a collection of military handbooks and field manuals that are essential contributions to the American outdoors tradition despite originating with the military. In the most basic sense, these books are the collection of experiences by the great experts of outdoors life: our countless expert soldiers who learned to thrive in the backwoods, deserts, extreme cold environments, and jungles of the world.

With technology playing a major role in everyday life, sometimes we need to take a step back in time to find those basic building blocks used for gaining mastery – the things that we have luckily not completely lost and has been recorded in books over the last two centuries. These skills aren't forgotten, they've just been shelved. *It's time to unshelve them once again and reclaim the lost knowledge of self-sufficiency.*

Based on this commitment to preserving our outdoors heritage, we have taken great pride in publishing this book as a complete original work. We hope it is worthy of both study and collection by outdoors folk in the modern era of outdoors and traditional skills life.

Unlike many other photocopy reproductions of classic books that are common on the market, this Historic Edition does not simply place poor photography of old texts on our pages and use error-prone optical scanning or computer-generated text. We want our work to speak for itself, and reflect the quality demanded by our customers who spend their hard-earned money. With this in mind, each Historic Edition book that has been chosen for publication is carefully remastered from original print books, *with the Doublebit Civilian Reference Edition printed and laid out in the exact way that it was presented at its original publication*. We provide a beautiful, memorable experience that is as true to the original text as best as possible, but with the aid of modern technology to make as beautiful a reading experience as possible for books that are typically over a century old. Military historians and outdoors enthusiasts alike are sure to appreciate the care to preserve this work!

Because of its age and because it is presented in its original form, the book may contain misspellings, inking errors, and other print blemishes that were common for the age. However, these are exactly the things that we feel give the book its character, which we preserved in this Historic Edition. During digitization, we ensured that each illustration in the text was clean and sharp with the least amount of loss from being copied and digitized as possible. Full-page plate illustrations are presented as they were found, often including the extra blank page that was often behind a plate. For the covers, we use the original cover design to give the book its original feel. We are sure you'll appreciate the fine touches and attention to detail that your Historic Edition has to offer.

For outdoors and military history enthusiasts who demand the best from their equipment, the Doublebit Press Civilian Reference Edition reprint of this military manual was made with you in mind. Both important and minor details have equally both been accounted for by our publishing staff, down to the cover, font, layout, and images. It is the goal of Doublebit Civilian Reference Edition series to preserve outdoors heritage, but also be cherished as collectible pieces, worthy of collection in any outdoorsperson's library and that can be passed to future generations.

*FM 20-15

FIELD MANUAL
No. 20-15

DEPARTMENT OF THE ARMY
WASHINGTON 25, D. C., *9 January 1956*

TENTS AND TENT PITCHING

	Paragraph	Page
CHAPTER 1. INTRODUCTION	1, 2	3
2. TENT AND TENT FLIES		
Section I. Arctic 10-man tent	3–8	3
II. Assembly tent	9–14	9
III. Command post tent	15–20	17
IV. General purpose tent, large	21–26	25
V. General purpose tent, medium	27–32	31
VI. Hexagonal tent	33–38	41
VII. Hospital tent, sectional	39–44	46
VIII. Kitchen tent	45–50	54
IX. Latrine screen	51–56	61
X. Maintenance shelter tent	57–61	65
XI. Mountain tent	62–66	72
XII. Red Cross markers	67, 68	76
XIII. Wall tent, large	69–74	77
XIV. Wall tent, small	75–80	85
CHAPTER 3. PINS, POLES, AND LINES	81–83	88
4. SITE SELECTION	84–86	93
5. HEATING AND VENTILATION	87, 88	96
6. GROUND COVERINGS	89–92	98
7. CARE OF TENTAGE		
Section I. Protection of tents against damage	93–97	99
II. Protection of pins, poles, and lines against damage	98–100	101
III. Repair methods	101–103	101
APPENDIX I. REFERENCES		103
II. GLOSSARY		104
INDEX		106

*This manual supersedes FM 20-15, 24 February 1945, including C 1, 23 December 1947; C 2, 16 January 1951; and C 3, 20 August 1951.

For sale by the Superintendent of Documents, U. S. Government Printing Office
Washington 25, D. C. — Price 55 cents

CHAPTER 1

INTRODUCTION

1. Purpose and Scope

a. The purpose of this manual is to provide information on the care and handling of tents issued by the Army. Most of the commonly used standard tents are included in the manual. For information on the M-1948, 16- by 16-foot, sectional, insulated, frame-type tent, see TM 10-616.

b. The manual is designed to serve as an aid for training personnel in the use of tents as well as a handy reference and guide in the field.

2. Basis for Tent Issue

The basis for the issue of tents may be found in—

a. TA 20—Field Installation and Activities. Tentage authorized for posts, camps, and stations.

b. TA 21—Clothing and Equipment. Tentage authorized for individuals.

c. TOE—Tentage authorized for units.

CHAPTER 2

TENT AND TENT FLIES

Section I. ARCTIC 10-MAN TENT

3. Use

The tent, arctic, 10-man, FWWMR, OD, complete with pins and poles (fig. 1), is used to provide shelter for troops operating in extremely cold and cold-wet areas. The tent normally accommodates 10 men and their individual clothing and equipment. It may also be used as a command post tent or as a small storage tent.

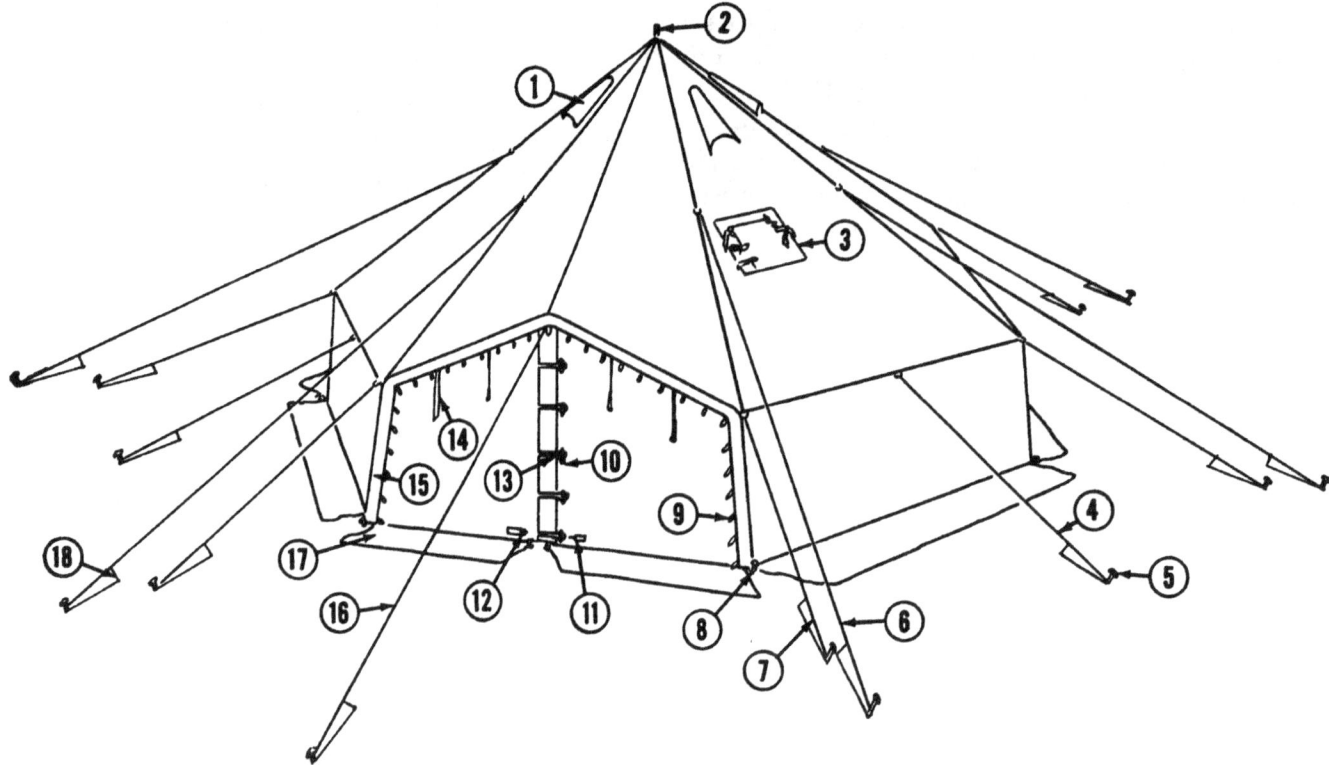

Reference No.	Part	Quantity	Federal Stock No.
1	Ventilator	4	
2	Pole, tent, telescopic, upright, jointed, magnesium, 4' 9" to 9' long	1	8340-188-8413
3	Stovepipe opening	1	
4	Line, tent, cotton, 12' 6", unfinished, 7/32" dia. (intermediate eave line)	4	8340-252-6912
5	Pin, tent, aluminum, 9"	28	8340-261-9749
6	Line, tent, 20', unfinished, 7/32" dia. (corner line)	6	No stock number available
7	Line, tent, cotton, 12' 6", unfinished, 7/32" dia., yellow (corner eave line)	6	8340-262-3658
8	Line, tent, cotton, footstop, 19" long, 1/4" dia. (footstop) (four used for screen doors)	14	8340-252-2299
9	Becket	62	
10	Toggle, tent, wood	10	8340-376-1295
11	Snap chape (outside)	2	
12	D-ring (outside)	2	
13	Loop, toggle	10	
14	Tie tape	8	
15	Lug (double)	2	
16	Line, tent, cotton, 21' 6", unfinished, 7/32" dia. (door eave line)	2	8340-252-6913
17	Snow cloth		
18	Slip, tent, line, magnesium	18	8340-223-8095
	Line, tent, 18' 9", unfinished, 1/8" dia. (sock line)	2	No stock number available
	Line, tent, 28' 6", unfinished, 1/8" dia. (sock line)	1	No stock number available
	Line, tent, 40' 6", unfinished, 1/8" dia. (sock line)	1	No stock number available
	Toggle, tent, steel, 1¾"	48	8340-242-7872
	Snap chape (inside)	2	
	D-ring (inside)	2	
	Door, screen, for tent, arctic, 10-man	2	8340-377-6950
	Tent liner, FWMR, arctic, 10-man	1	8340-262-3698
	Cover, tent, FWWMR, arctic, 10-man	1	8340-262-3653

Figure 1. Tent, arctic, 10-man, FWWMR, OD, complete with pins and poles, Federal Stock No. 8340-262-3685.

4. Description

The tent is a six-sided pyramidal tent, supported by a telescopic center pole.

a. Dimensions. Each side of the tent is 8 feet 9 inches long. The tent is 8 feet 6 inches high at the peak. The wall height is 3 feet, giving a pitch of 5 feet 9 inches. The hexagonal floor of the tent is 17 feet 6 inches in diameter.

b. Weight and Cubage. The tent weighs 68 pounds, and the pins and pole weigh 8 pounds. The tent has a cubage of 7.1 cubic feet, the pins and pole a cubage of 0.2 cubic feet.

c. Floorspace. The floorspace is 198.9 square feet.

d. Material. The tent is made of 9-ounce olive-drab sateen cotton cloth, FWWMR.

e. Doors. The tent has two doors 5 feet high on opposite sides, permitting tents to be joined together with suitable access from one to the other. Door flaps may be securely closed either by slide fasteners or by loops over wood toggles. The doors are operated from both inside and outside.

f. Ventilation. The tent is ventilated by 4 built-in ventilators on opposite sides and near the peak of the tent. The ventilators have inside ducts, which may be closed by tie cords. The ventilator hoods are of the fixed type, each hood being constructed with a stiffener inserted in the hem to keep it extended out from the ventilator opening.

g. Heating. The tent is heated by an M-1950 Yukon stove. A stovepipe opening with a silicone rubber-molded ring is built in one side of the tent near the eave. When the stove is not in use, the stovepipe opening may be protected by a canvas flap.

h. Snow Cloths. There is a snow cloth sewed to the bottom of each side of the tent. When the tent is pitched, the snow cloths are flat on the ground on the outside of the tent. Snow is deposited on the snow cloths for insulation purposes.

i. Screen Doors. Two screen doors are provided; they may be attached to the front and rear of the tent for protection against insects.

j. Sock Lines. Four sock lines are provided for drying clothing and equipment.

k. Liner. A fire-resistant liner, made of 5.2-ounce permeable cotton sheeting, is provided to insulate the tent and to prevent frost from falling on the occupants. The liner is held in place by metal toggles.

l. Cover. The tent is provided with a cover for use when in storage or being transported. The tent and liner, when folded, fit into the cover. Aluminum tent pins are nested and the magnesium pole telescoped to its shortest length and placed in the pocket at one side of the cover.

5. Ground Plan

Before pitching the tent, study the ground plan carefully (fig. 2).

6. Pitching

The tent can be pitched by 6 men in approximately 27 minutes.

a. Preliminary Procedures (1, fig. 3).
  (1) Spread tent on ground. Check to see if liner is in place; usually, it is not in place in a new tent. If liner is not in place, spread it out beneath the tent.
  (2) Secure **D**-rings to snaps inside front and rear doors.
  (3) Close slide fasteners in front and rear doors.
  (4) Secure **D**-rings to snaps outside front and rear doors.
  (5) Drive 6 corner pins and 4 door pins and attach footstops to pins.

b. Attaching Corner Eave Lines and Inserting Tentpole (2, fig. 3).
  (1) Drive 6 pins about 7 feet from corners of tent and attach yellow corner eave lines. Pins on opposite sides of tent should be in a straight line.
  (2) Open front door and push pole, extended to 8 feet 6 inches, under tent.
  (3) Insert spindle of pole through hole in peak of liner and through supporting ring in peak of the tent.

c. Raising Tent (3, fig. 3).
  (1) With one man inside the tent, close inside and outside **D**-rings and snaps on doors; close slide fasteners.
  (2) Fasten loops to wood toggles on doors.
  (3) Lift tentpole, and line up door openings, stovepipe, and four vent openings of liner with openings in tent.
  (4) Insert **D**-rings of liner into snaps attached to tent.
  (5) Raise tentpole, placing butt of tentpole in center of tent area.

d. Attaching Door Eave Lines, Intermediate Eave Lines, and Corner Lines (4, fig. 3).

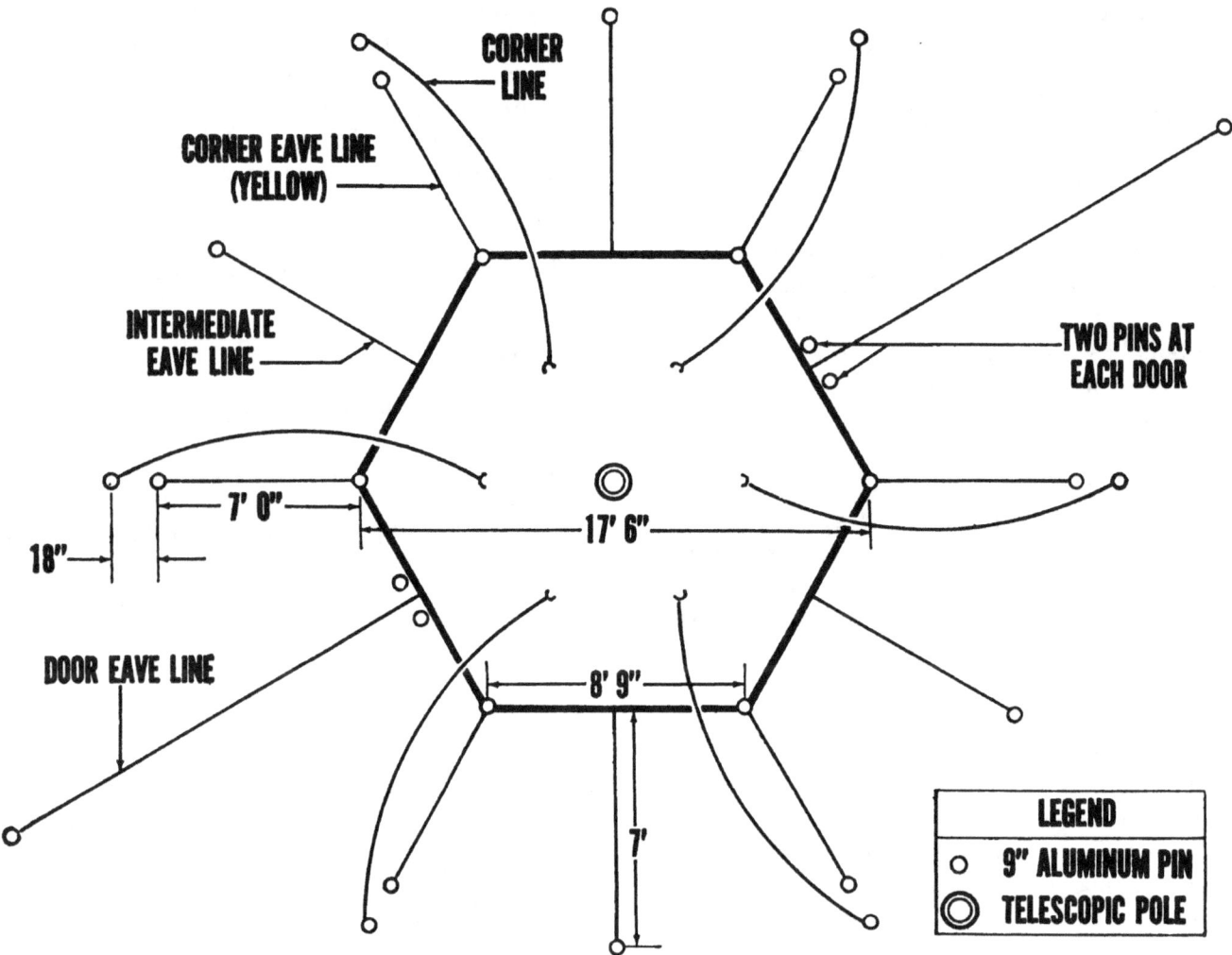

Figure 2. Ground plan of tent, arctic, 10-man, FWWMR, OD.

(1) Stake the 2 door eave lines far enough to hold doors vertical.

(2) Attach the 4 intermediate eave lines to pins.

(3) Attach the 6 corner lines to pins 18 inches out from corner eave-line pins.

(4) Adjust and tighten all lines.

e. Propping Up Door Eave Lines. Each of the 2 door eave lines may be propped up by placing the line over an improvised pole (tree branch or other object higher than the door entrance) at a distance of about 5 feet in front of the door and then staking the line out to a pin. This keeps the doors from sagging, makes the slide fastener work better, makes the tent easier to get into and out of, and gives the tent greater stability.

f. Fastening Liner.

(1) Insert metal toggles through grommets of liner, allowing approximately 2 inches between tent and liner for insulating purposes.

(2) Tie tapes around stovepipe opening in liner to corresponding tapes around stovepipe opening in tent to keep stovepipe opening in place.

(3) Tie one end of the 18-foot 9-inch sock line to toggle in each corner of door, threading line through eye of toggles at eave line and tying to carrier toggles of the opposite door. Use same procedure for the 18-foot 9-inch sock line on opposite side of tent.

(4) Thread the 40-foot 6-inch sock line through the next line of toggles, encircling the tent, and tie.

(5) Secure the 38-foot 6-inch sock line in like manner in the next row of toggles.

g. Joining Two Tents Together.

(1) When two tents are to be joined together,

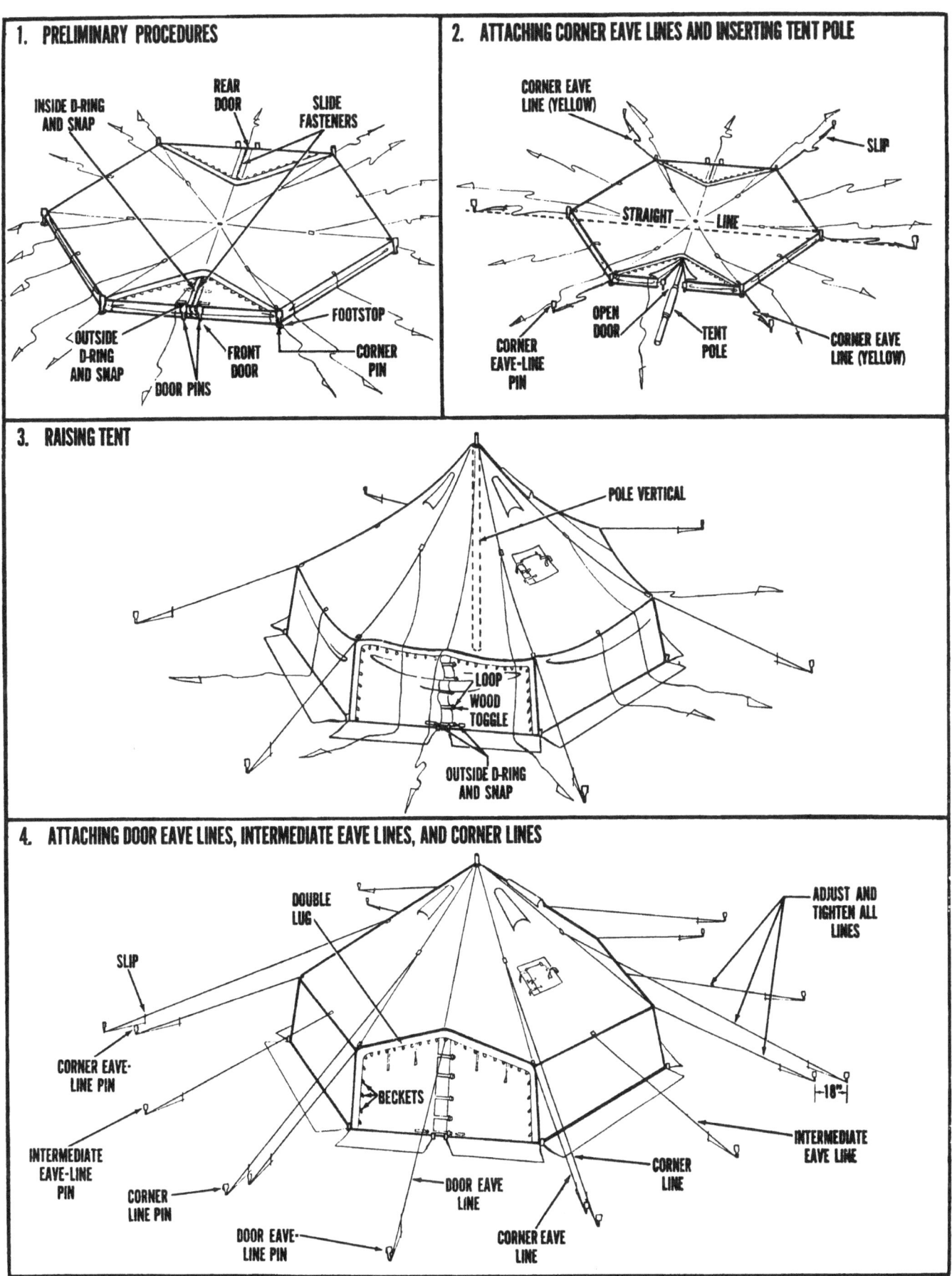

Figure 3. Steps in pitching tent, arctic, 10-man, FWWMR, OD.

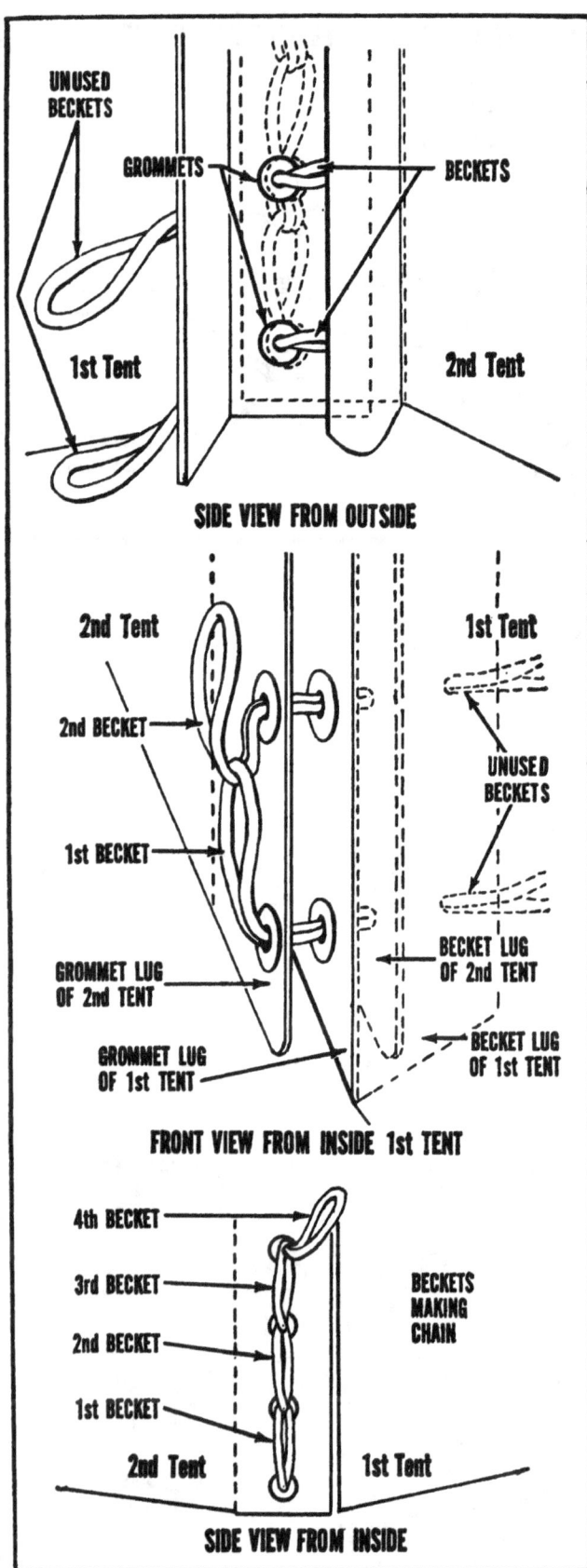

Figure 4. Joining two tents together by chain-lacing beckets through grommets (tent, arctic, 10-man, FWWMR, OD.

erect the first tent as described above. Fasten lugs (4, fig. 3) at front or rear of tents together by inserting grommet lug of one tent between grommet lug and the becket lug of other tent and chain-lace beckets (4, fig. 3) on lug of one tent through grommets on each lug of both tents (fig. 4). Begin chain lacing at bottom (near the ground) of lugs and continue until bottom (near the ground) at the other end of the same lugs is reached, securing last becket with a knot. Then erect second tent in the same manner as first tent.

(2) An alternate method of joining two tents together is to spread both tents on the ground with the front or rear of one tent next to the front or rear of the other tent and fasten the lugs of the two tents together as in (1) above. Then erect the two tents as described above.

h. Attaching Screen Doors. Attach screen doors, if not already stitched to tent, by chain-lacing beckets on lugs at front and rear of tent through grommet lugs of tent and screen doors. Chain-lace screen doors to lugs of tent in the same manner as when two tents are joined together (*g* above).

7. Striking

a. Remove screen doors.

b. Remove door eave lines from pins.

c. Loosen footstops from pins and remove footstop pins.

d. Loosen all other lines and remove all other pins.

e. Remove tentpole, and telescope pole to its shortest length.

f. Remove liner only if repairs are needed.

8. Folding

a. Folding Tent.

(1) *Engaging snaps into D-rings and closing slide fasteners.* Engage snap into D-ring inside doors, and close door slide fasteners.

(2) *Spreading tent out to fold* (1, fig. 5). Spread tent on ground and locate stovepipe opening panel. Grasp corner eave line (to right of stovepipe opening) and pull out corner of panel. Then coil inter-

mediate eave line neatly on extended panel.

(3) *Making first panel fold* (2, fig. 5). Reaching to the left, grasp corner eave line (to left of stovepipe opening) and pull second panel to the right, making an accordion fold.

(4) *Folding remaining panels* (3, fig. 5). Fold remaining panels in the same manner, having 6 folds in all. As each fold is completed, coil intermediate eave lines or door eave lines neatly between folds.

(5) *Coiling lines on top of folded panels* (4 fig. 5). Coil on top of folded tent panels the 6 corner lines and the 6 corner eave lines that have been drawn to the right and the last remaining intermediate eave line.

(6) *Folding tent peak down 4 feet* (5, fig. 5). Grasp peak of tent and fold so that peak extends down tent deck approximately 4 feet. Fold snow cloth up over side walls of tent.

(7) *Folding tent over* (6, fig. 5). Fold tent approximately in half along its long dimension.

(8) *Folding tent toward center and placing it on cover with screen doors, pins, and pole* (7, fig. 5). Fold edges of tent toward center so that no portion of liner is exposed. Place folded tent on cover; place folded screen doors on top of folded tent; place nested pins and telescoped pole into pocket of cover.

(9) *Closing cover*. Close cover, securing it with straps and loops. Care should be taken that no portion of the tent is exposed and that the flaps are tucked neatly within the cover.

b. *Folding Liner*. Ordinarily, the liner is not removed from the tent. When necessary, the liner may be folded separately in the same manner as the tent and placed inside the cover with the tent, screen doors, pins, and pole.

Section II. ASSEMBLY TENT

9. Use

The tent, assembly, M-1942, FWWMR, OD, complete with pins and poles (fig. 6), is used for church services in the field, for lectures, and for the showing of movies. It may also be used for storage, for quartering personnel, or for any other authorized purpose. It has a seating capacity of approximately 500 men. When used for quartering personnel, it has a capacity of approximately 80 men.

10. Description

The tent is a large special-purpose tent, with a rectangular middle section and rounded hip-roofed ends. The top is made in 4 sections which lace together: two middle sections and two rounded end sections. The side wall is in 4 sections. There are 3 chains and supporting rings and 3 sets of block and tackle with lines. Since the tent is sectional in construction, it may be extended to any desired length by means of additional middle and wall sections. Two end sections can be joined and used with two wall sections to form a circular tent.

a. *Dimensions*. The tent is 40 feet wide, 80 feet long and 18 feet high. The 3 main center poles are 21 feet in length and the wall is 8 feet high. When the peaks of the tent are set by block and tackle, the height from the ground to the top of the tent ridge is 18 feet, with an additional 3 feet of the main poles protruding above the top to accommodate the block and tackle. This gives the tent a pitch of 10 feet.

b. *Weight and Cubage*. The tent weighs 1,100 pounds, and the pins and poles weigh 655 pounds. The tent has a cubage in storage of 23.3 cubic feet, the pins and poles a cubage of 16.9 cubic feet.

c. *Floorspace*. The floorspace is 2,856.6 square feet.

d. *Material*. The middle and end sections are made of 12.29-ounce duck, FWWMR. The wall sections are made of 9.85-ounce duck.

e. *Door Openings*. The tent has 4 door openings, each of which is made by the overlapping of a panel and a half of side wall where two sections of the side wall meet. The side walls may be shifted so that the openings come anywhere as long as the same proportionate distance between the openings is maintained.

f. *Ventilation*. There are 4 built-in ventilators, one to each side wall section. The tent may also be ventilated by rolling up the side walls and tying

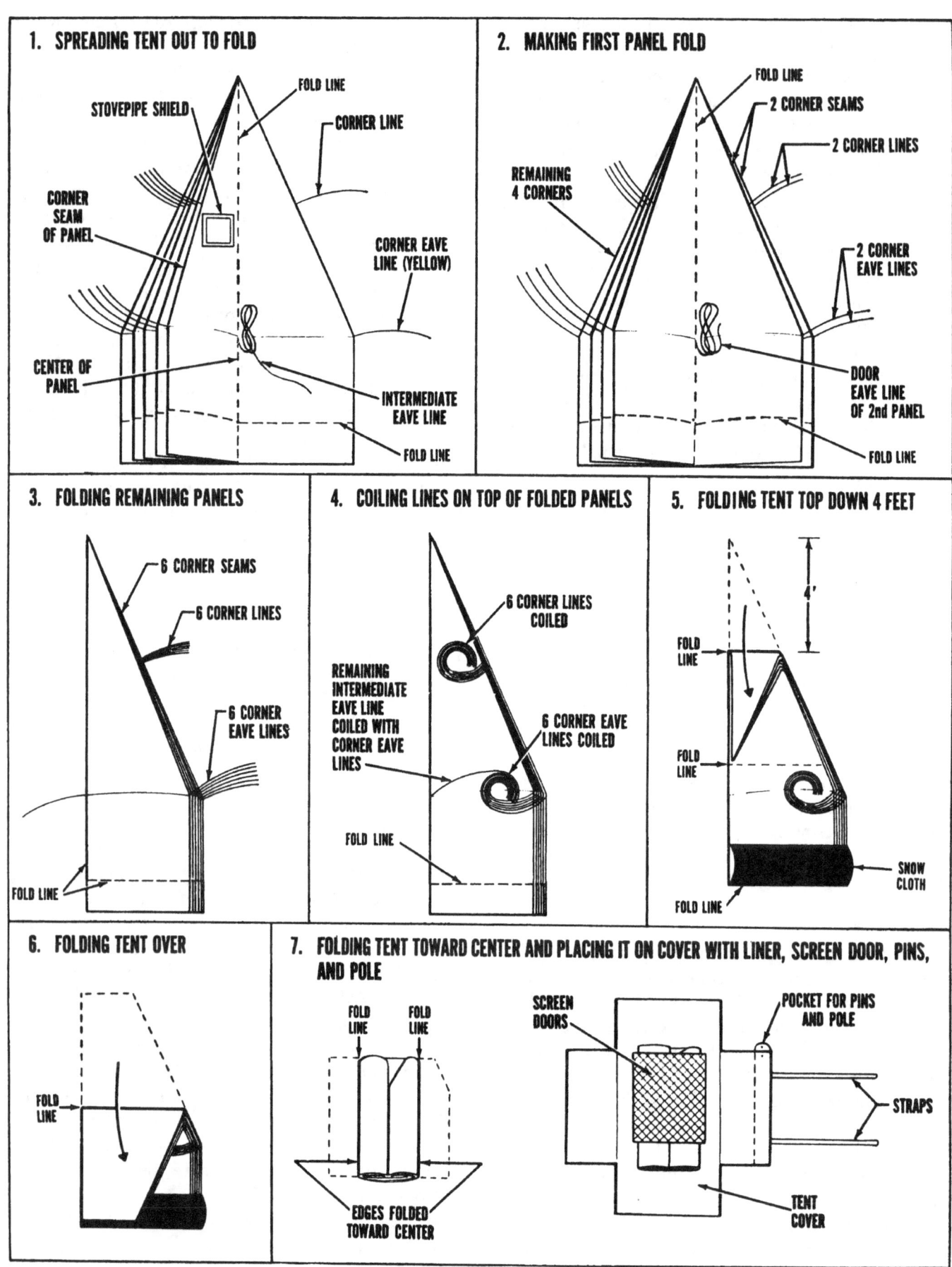

Figure 5. Steps in folding tent, arctic, 10-man, FWWMR, OD.

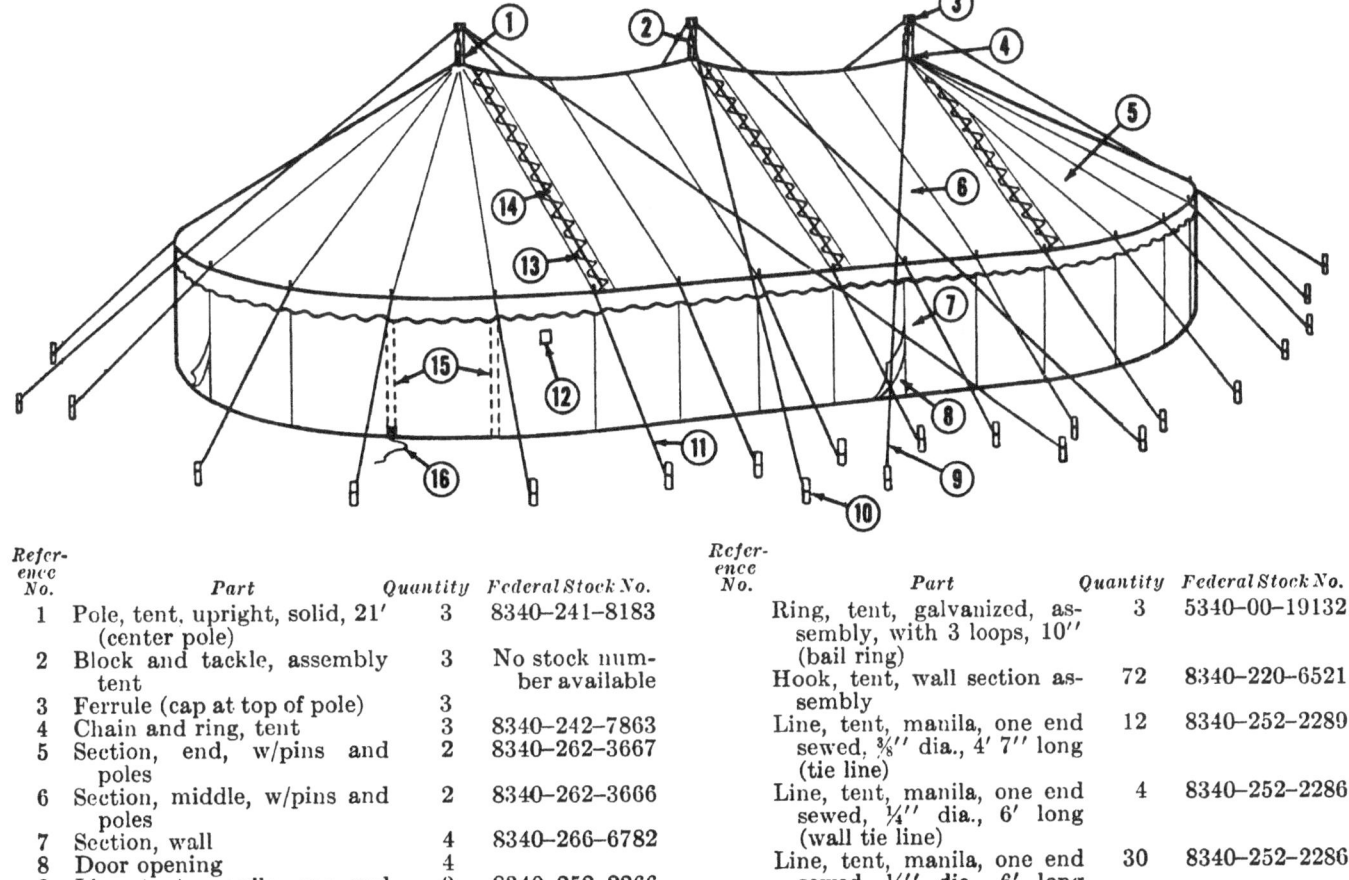

Reference No.	Part	Quantity	Federal Stock No.
1	Pole, tent, upright, solid, 21' (center pole)	3	8340-241-8183
2	Block and tackle, assembly tent	3	No stock number available
3	Ferrule (cap at top of pole)	3	
4	Chain and ring, tent	3	8340-242-7863
5	Section, end, w/pins and poles	2	8340-262-3667
6	Section, middle, w/pins and poles	2	8340-262-3666
7	Section, wall	4	8340-266-6782
8	Door opening	4	
9	Line, tent, manila, one end sewed, one end w/thimble and hook, ½" dia., 52' long (guy line on 21' pole)	9	8340-252-2266
10	Pin, tent, wood, 36"	39	8340-261-9752
11	Line, tent, manila, one end sewed, ½" dia., 18' 6" long (eave line)	30	8340-252-2278
12	Stovepipe opening	4	
13	Line, tent, manila, one end sewed, ¼" dia., 27' long (lacing line, extension cloth)	6	8340-252-2277
14	Extension cloth	6	
15	Pole, tent, upright, solid, 8' 3" (eave pole)	30	8340-188-8409
16	Line, tent, manila, one end sewed, ¼" dia., 3' 4" long (wall tie line)	36	8340-252-2291
	Ring, tent, galvanized, assembly, with 3 loops, 10" (bail ring)	3	5340-00-19132
	Hook, tent, wall section assembly	72	8340-220-6521
	Line, tent, manila, one end sewed, ⅜" dia., 4' 7" long (tie line)	12	8340-252-2289
	Line, tent, manila, one end sewed, ¼" dia., 6' long (wall tie line)	4	8340-252-2286
	Line, tent, manila, one end sewed, ¼" dia., 6' long (jumper line)	30	8340-252-2286
	Line, tent, manila, one end sewed, ½" dia., 6' long (lacing line, neck, center section)	4	8340-252-2287
	Line, tent, manila, both ends sewed, ½" dia., 12' long (lacing line, neck, end section)	2	8340-252-2294
	Thimble, tent, iron, 1¾"	52	4030-262-1785
	Thimble, tent, iron, 2¼"	24	4030-262-1787
	Thimble, tent, iron, 3"	4	4030-262-1789
	Cover, tent, assembly, M-1942	6	
	Line, tent, manila, one end sewed, one end w/eye, 5/16" dia., 13' long (cover tie line)	12	8340-252-2271

Figure 6. Tent, assembly, M-1942, FWWMR, OD, complete with pins and poles.

them with the attached tie tapes; or, if it is raining, by extending the side wall and tying it to the lines running from the eaves. The doors may also be tied back and the openings used for ventilation.

g. Heating. Four M-1941 tent stoves or two external gasoline tent heaters, 250,000 B. t. u., are used to heat the tent. The built-in ventilators are used as stovepipe openings when M-1941 tent stoves are used.

h. Covers. The tent is provided with 6 covers for use when in storage or when being transported.

11. Ground Plan

Before pitching the tent, study the ground plan carefully (fig. 7).

12. Pitching

The tent can be pitched by 9 men in approximately 90 minutes.

a. Spotting Center Poles (1, fig. 8). Spot the 3 center poles according to ground plan and place a marker at each location. Drive marker in about 6 inches.

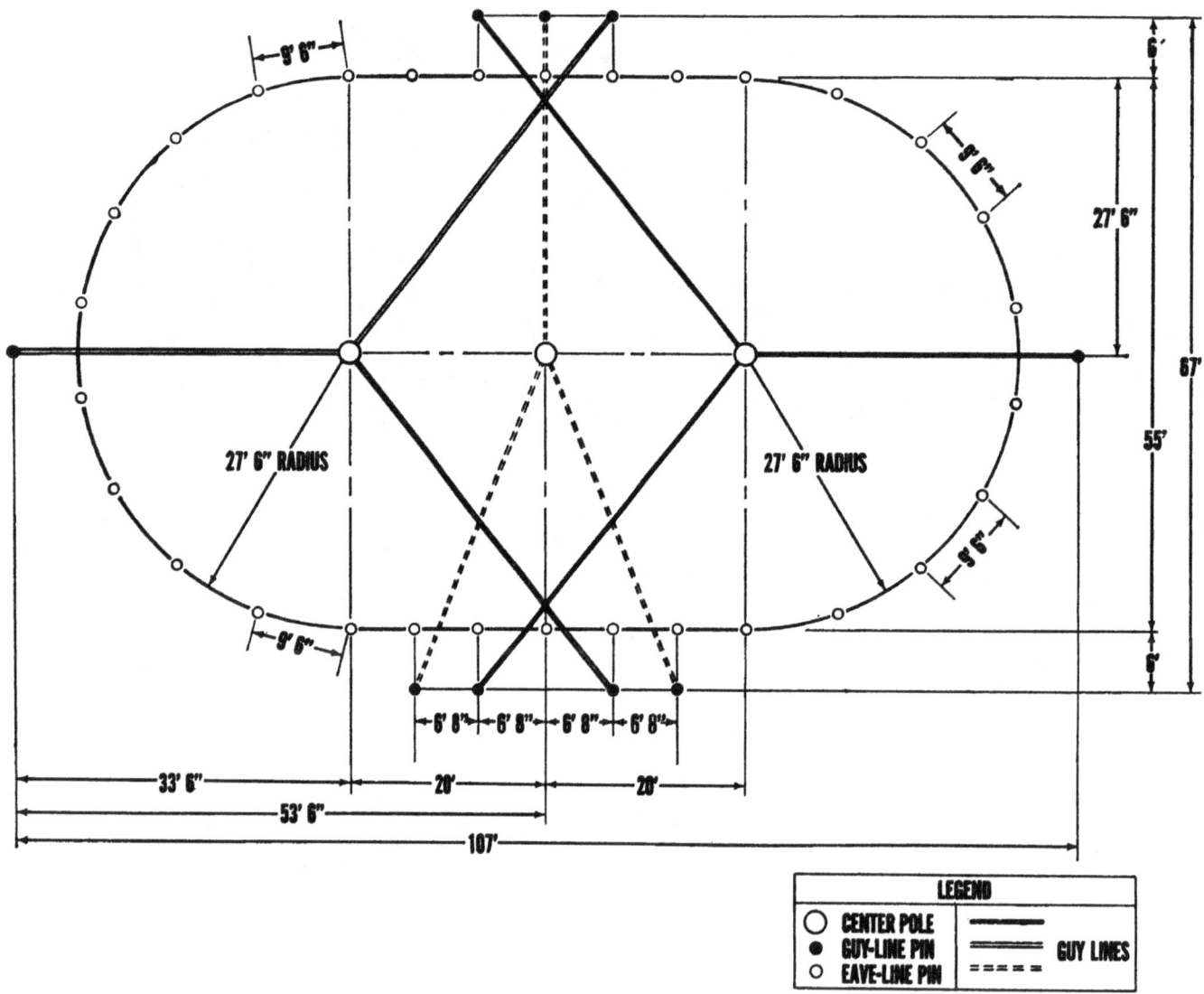

Figure 7. Ground plan of tent, assembly, M-1942, FWWMR, OD.

b. *Laying Out and Driving Pins* (2, fig. 8).

(1) *Eave-line pins.* Lay out and drive the 30 eave-line pins according to ground plan. Make sure that they are driven vertically and that the top of each pin is no more than 10 inches from the ground. The 27-foot extension cloth lacing line may be used as an aid in laying out the eave-line pins of the end sections.

(2) *Guy-line pins.* Lay out and drive guy-line pins according to ground plan. There are 9 guy-line pins, 3 for each center pole.

c. *Preparing Center Poles for Erection* (3, fig. 8 and fig. 9).

(1) Place the 3 center poles on the ground on one side of tent area. The poles should be perpendicular to the eave-line pins, and the butt end of each pole should be at a pole marker.

(2) Attach 3 main guy lines and one set of block and tackle to ferrule at top end of each pole. Lash drift line of block and tackle to pole, with single block 2 or 3 feet from butt end of pole. Place a bail ring assembly around butt end of each pole.

d. *Erecting Middle Center Pole* (4, fig. 8). One man stands at the butt end of the middle center pole, one man at the top end of the pole, and one man at the end of each of the 3 guy lines. One of these guy lines leads to the outside pin on a line at right angles from the center point of the long dimensional line of the tent layout (fig. 7). This places the man holding the line directly in line with the man at the butt end of the

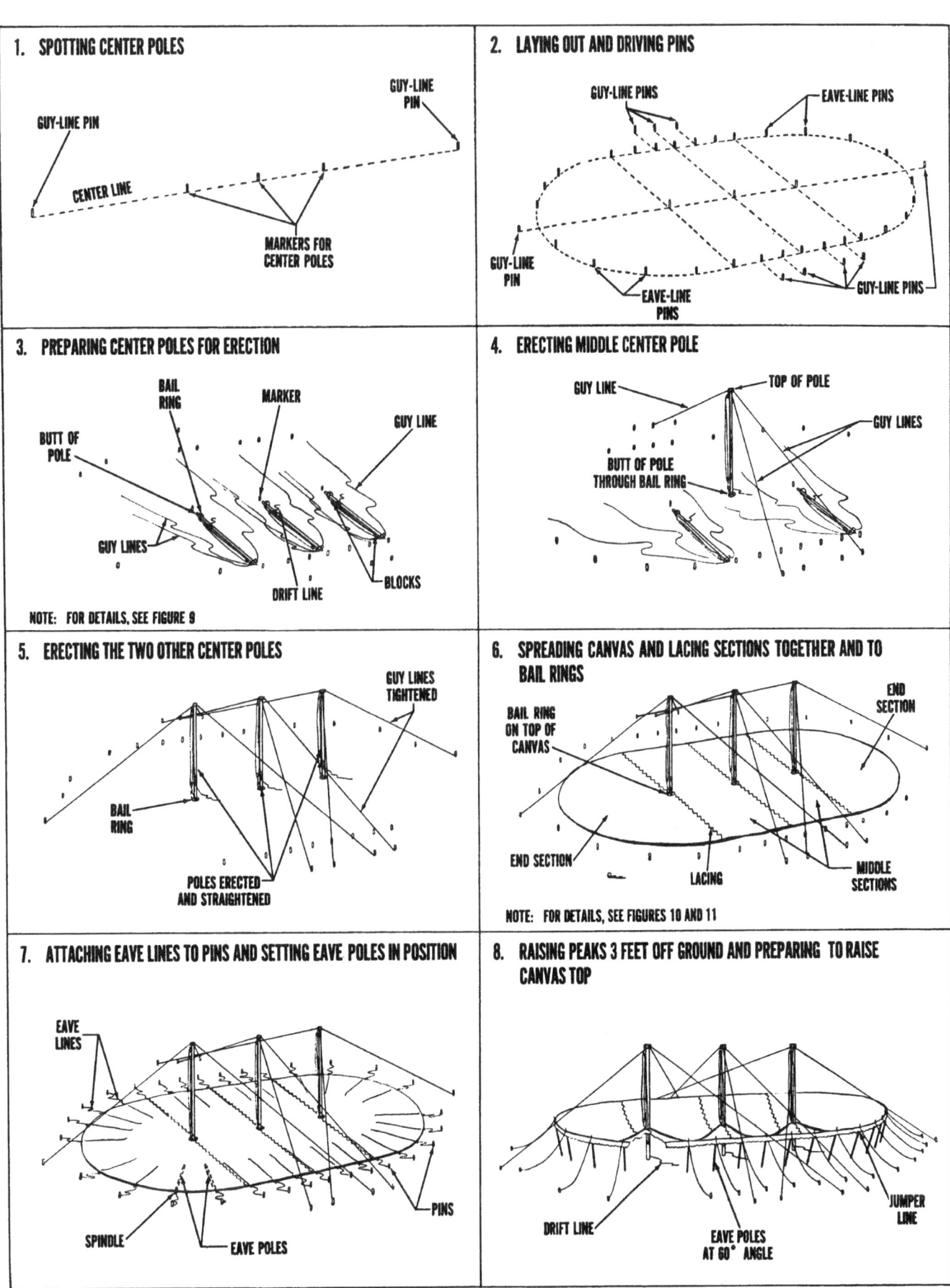

Figure 8. Steps in pitching tent, assembly, M-1942, FWWMR, OD.

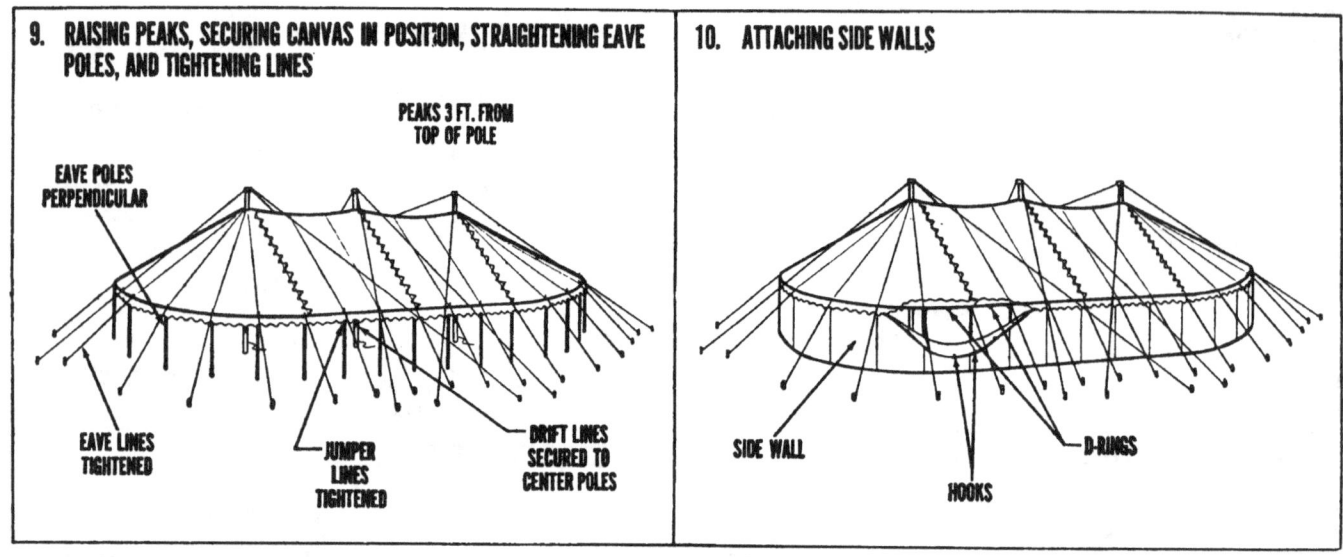

Figure 8—Continued.

pole. The man at the butt end of the pole keeps the pole in position with the marker by holding it with his foot. Be sure that the butt end of the pole is through the bail ring. The man at the top end raises the pole and walks toward the butt end. The man holding the center guy line assists by maintaining a taut line as the pole is raised. After the pole reaches such a height that it might swing off center, the men holding the other two guy lines spread out slowly to keep the pole balanced until it is in a perpendicular position. Then the guy lines are attached to the pins indicated on the ground plan.

e. Erecting the Other Two Center Poles (5, fig. 8). The other 2 center poles should be erected as in *d* above, except that one guy line leads to the outside pin on the direct center line of the long dimension of the tent layout. After poles have been erected, straighten them, and remove marker stakes. Tighten all center pole guy lines.

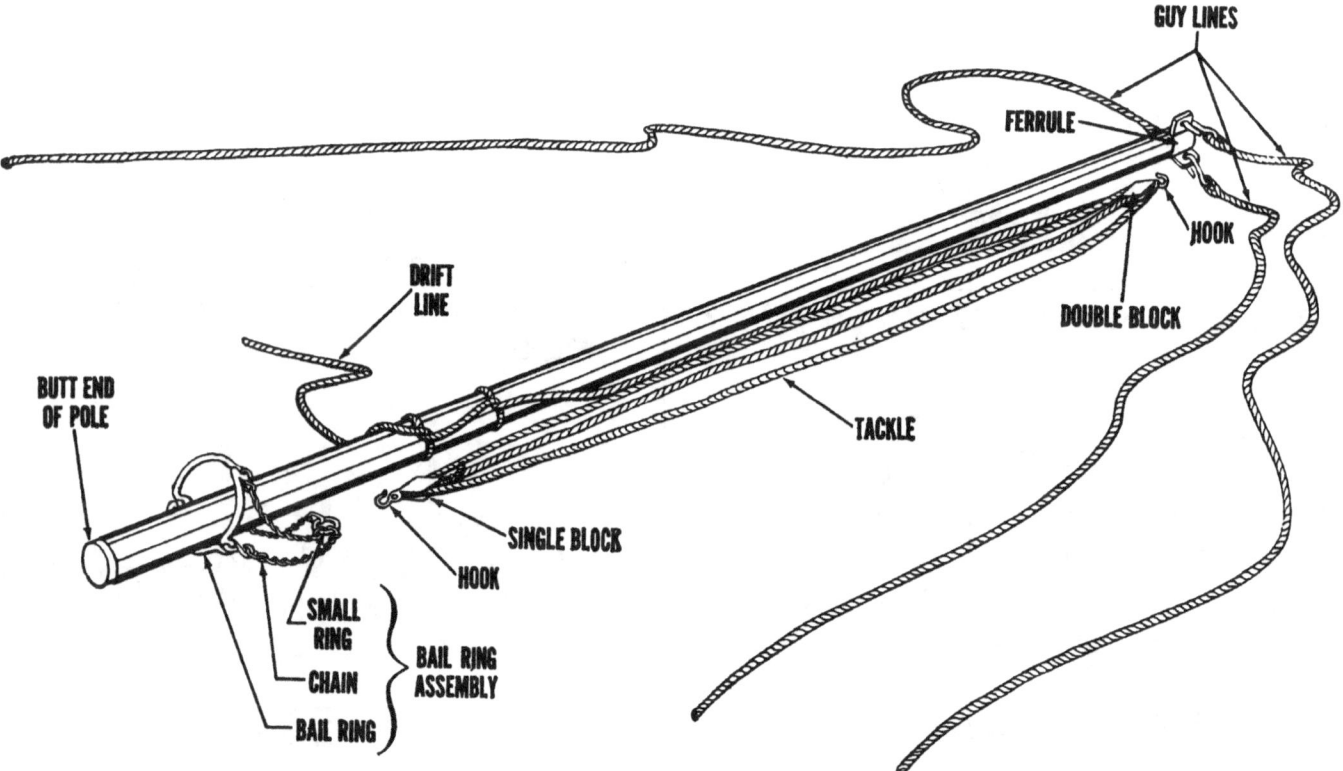

Figure 9. Preparing center pole for erection (tent, assembly, M-1942, FWWMR, OD).

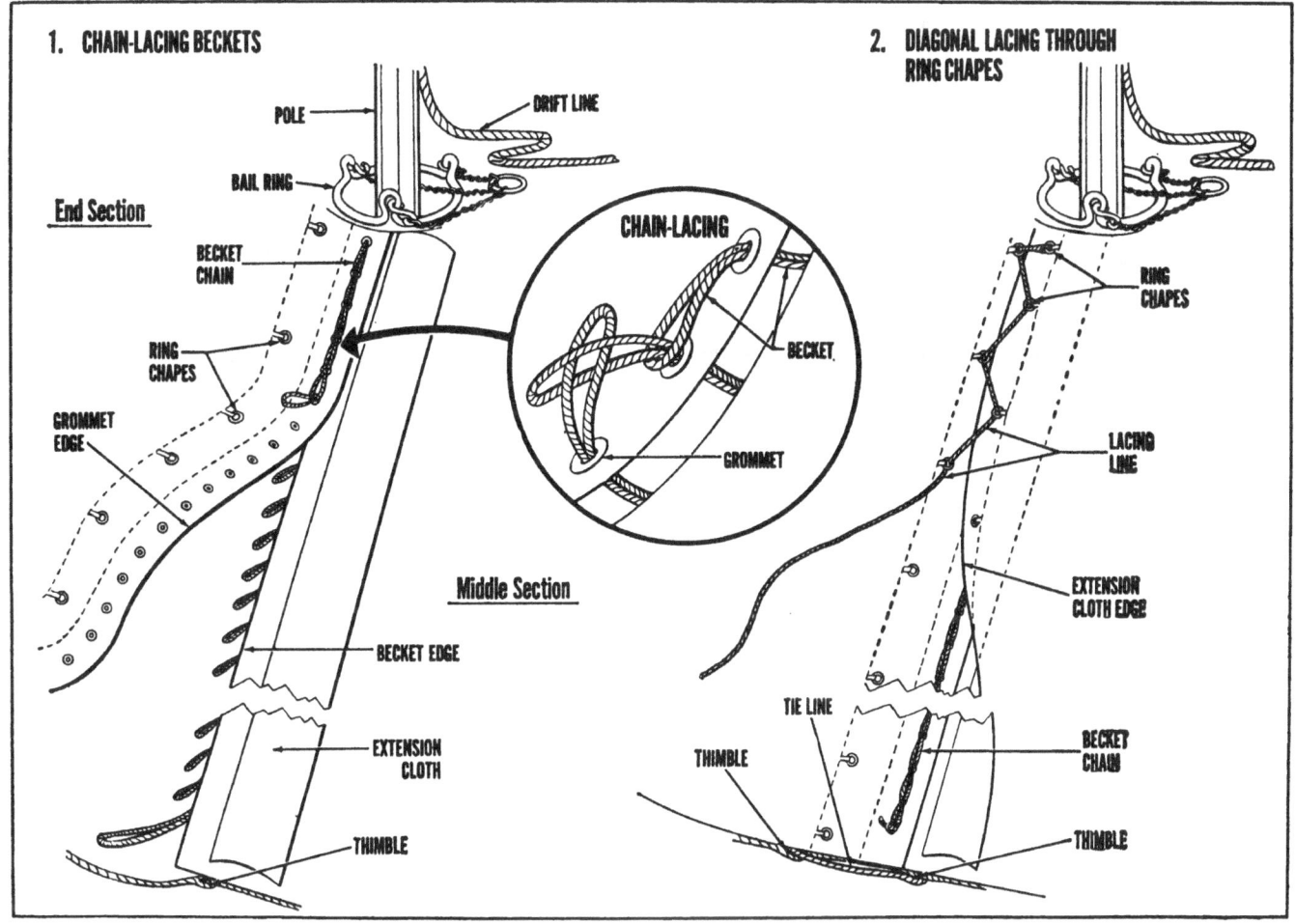

Figure 10. Steps in lacing top sections together (tent, assembly, M-1942, FWWMR, OD).

f. Spreading Canvas and Lacing Sections Together and to Bail Rings (6, fig. 8).

(1) Unfold the 2 middle sections and the 2 end sections. Spread sections on ground in position on tent area around the 3 center poles.

(2) Join sections from ridge to eave reinforcement line by chain-lacing beckets through grommets, securing the last becket through the last grommet with a knot (1, fig. 10).

(3) Secure eave corners of sections together by lashing tie line through thimble on eave corner of one section and through thimble on eave corner of the other section (2, fig. 10).

(4) Pull extension cloth over chain lacing, lace extension cloth lacing line diagonally through ring chapes, and tie end of line through eave corner thimbles (2, fig. 10).

(5) Attach hooks on single blocks to small rings of bail ring assemblies in order to lift assemblies off the ground about 1 foot (fig. 11).

(6) Secure sections of tent together at neck by lashing tie line on each side of neck through thimble of one section and through thimble of the other section (fig. 11).

(7) Fasten thimbles at necks of sections to bail rings by lacing neck lacing lines of two sections around bail ring and through thimbles (fig. 11).

(8) Unlash drift lines and put ends of lines through bail ring assemblies close to poles.

g. Attaching Eave Lines to Pins and Setting Eave Poles in Position (7, fig. 8).

(1) Attach, with 2 half hitches, all eave lines to pins approximately 2 feet in from the end of each line.

(2) Raise canvas at eave and slide butt end of eave poles toward a center pole. In-

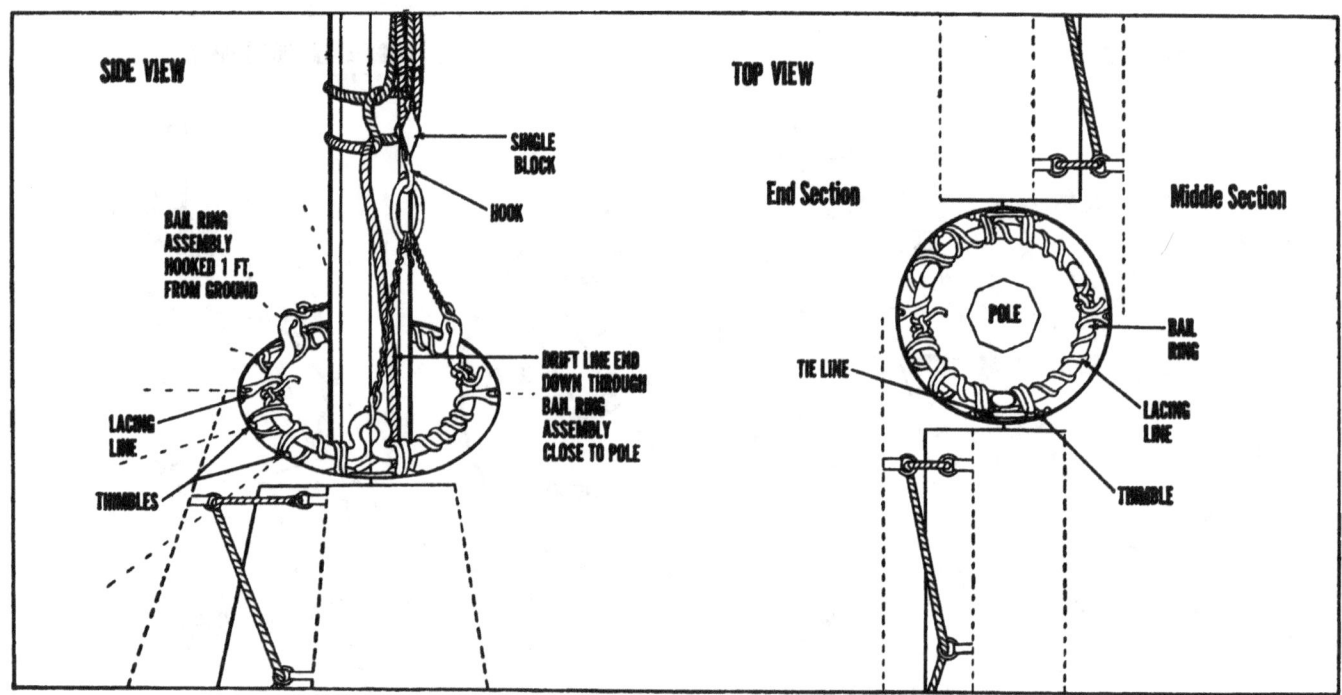

Figure 11. Lashing sections of top to bail ring around center pole (tent, assembly, M-1942, FWWMR, OD).

sert spindle of eave poles through leather reinforcements at point where eave lines are attached to canvas.

h. Raising Peaks 3 Feet Off Ground and Preparing to Raise Canvas Top (8, fig. 8).

(1) Going under canvas to center poles, raise peaks of tent about 3 feet off the ground by pulling drift lines. Lash drift lines to center poles, making sure that drift line of each block and tackle assembly is inside bail ring and next to center pole (fig. 11).

(2) Set eave poles to form an angle of about 60° with the ground, with butt of each pole pointing toward and in line with butt of nearest center pole.

(3) Fasten jumper line at eave of canvas to each eave pole with 2 half hitches.

(4) Partly tighten all eave lines.

i. Raising Peaks, Securing Canvas in Position, Straightening Eave Poles, and Tightening Lines (9, fig. 8).

(1) Raise peaks with drift line of block and tackle to within 3 feet of top of each center pole. The 3 peaks of the tent should be raised at the same time.

(2) Straighten all eave poles to a perpendicular position and tighten all lines as much as possible in order to eliminate wrinkles in tent roof. Lines are tightened or loosened by readjusting the 2 half hitches on each guy and eave line near the pin; there are no tent slips used with the assembly tent.

(3) Secure drift lines firmly to center poles.

j. Attaching Side Walls (10, fig. 8). Attach side walls by hooking wall hooks on top of side walls through D-rings attached to top sections of tent.

13. Striking

The tent can be struck by 9 men in approximately 60 minutes.

a. Checking Center Pole Guy Lines. Check center pole guy lines, making sure that they are hooked in ferrule at top of center pole and are taut.

b. Detaching Side Walls. Detach side walls by unhooking wall hooks from D-rings.

c. Adjusting Eave Poles. Slant butts of eave poles toward butts of center poles at a 60° angle with the ground. If weather is calm, untie eave pole jumper lines; do not untie jumper lines in a high wind.

d. Letting Down Peaks. Let peaks down to ground level by releasing drift lines, making sure that entire canvas area is in such a position that when sections are unlaced, there will be little difficulty in folding them.

e. Removing Eave Poles. Remove all eave poles.

f. Removing Pins. Remove all eave line pins.

g. Unlacing Sections. Unlace sections of tent, remove from beneath center poles, and separate for folding into separate bundles.

h. Striking Center Poles. Strike center poles. To strike a center pole, 2 men stand at butt end of pole and 1 man at end of each guy line. Untie guy lines from pins. The men holding the ends of the guy lines then walk slowly toward the center of the tent area, keeping the lines taut to prevent the pole from swaying. One of the men at the butt end of the pole walks slowly forward with the pole, easing it gradually to the ground, while the other man at the butt end of the pole steadies it.

14. Folding

The tent can be folded and placed into 6 covers by 9 men in approximately 20 minutes.

a. Folding Middle Sections (1, fig. 12). Fold each of the 2 middle sections in half along the long dimensions and then in half again. Then, in 2½-foot folds, fold ends toward center.

b. Folding End Sections (2, fig. 12). Fold each of the 2 end sections in half along the long dimension and then in half again. Then, in 2½-foot folds, fold ends toward center.

c. Folding Wall Sections (3, fig. 12). Fold each of the 4 wall sections in half along the long dimension. Then, in 2½-foot folds, fold ends toward center.

d. Putting Folded Sections Into 6 Separate Covers (4, fig. 12). Put folded sections into 6 separate covers. Place each middle and end section in a separate cover; place 2 wall sections in a separate cover. Fold flaps of each cover over folded section and tie through the grommets the 2 cover tie lines provided.

Section III. COMMAND POST TENT

15. Use

The tent, command post, M-1945, FWWMR, OD, complete with pins and poles (fig. 13), is used in theaters of operations to provide office space for staff sections, accommodating 3 men and the necessary folding tables and office equipment. It may also be used for the quartering of 3 officers or as a battalion aid station, the blackout vestibule being long enough to accommodate a litter and bearers.

16. Description

The central part of the tent is A-shaped. The ends are hip-roofed with converging side walls.

a. Dimensions. The tent is 10 feet wide, 20 feet 7 inches long (6 ft 10 in. of which is vestibule), and 9 feet high. The side wall height is 5 feet 6 inches, giving a pitch of 3 feet 6 inches.

b. Weight and Cubage. The tent weighs 165 pounds, and the pins and poles weigh 92 pounds. The tent has a cubage in storage of 6.3 cubic feet, the pins and poles a cubage of 3.6 cubic feet.

c. Floorspace. The floorspace is 172 square feet, of which 48 square feet is vestibule space.

d. Material. The tent is made of 12.29-ounce duck, FWWMR. The canvas is supported on a webbing framework, which carries the weight of the canvas. Fair-leads carry the stress between webbing and eave lines and eliminate friction between eave and eave lines. The tent walls, tent top, and sod cloth are constructed of one piece.

e. Door. The tent has a door entrance at the front, 6 feet high and 4 feet wide.

f. Blackout Curtain. A nondetachable blackout curtain, with a slide fastener opening (5, fig. 15), is sewed into the body of the tent. The curtain separates the vestibule from the main part of the tent. When the tent is used for a first-aid station, the vestibule space between the door and the blackout curtain is large enough to allow stretcher bearers passage without emitting light.

g. Windows. The tent has three 24-inch square window sashes, made of flexible translucent material. The sashes are inserted in window openings and held in place by snap fasteners. Canvas flaps cover the window during blackouts.

h. Ventilation. The tent is ventilated by an opening near the top of the rear end section. The ventilator has an inside duct, which may be closed by a tie cord. The ventilator hood is of the fixed type, constructed with a stiffener inserted in the hem to keep it extended out from the ventilator opening. For additional ventilation, the side walls can be rolled up and the side wall screens attached.

i. Heating. The tent is heated by an M-1941 tent stove. There is a stovepipe opening built in the top of the tent near the rear center upright

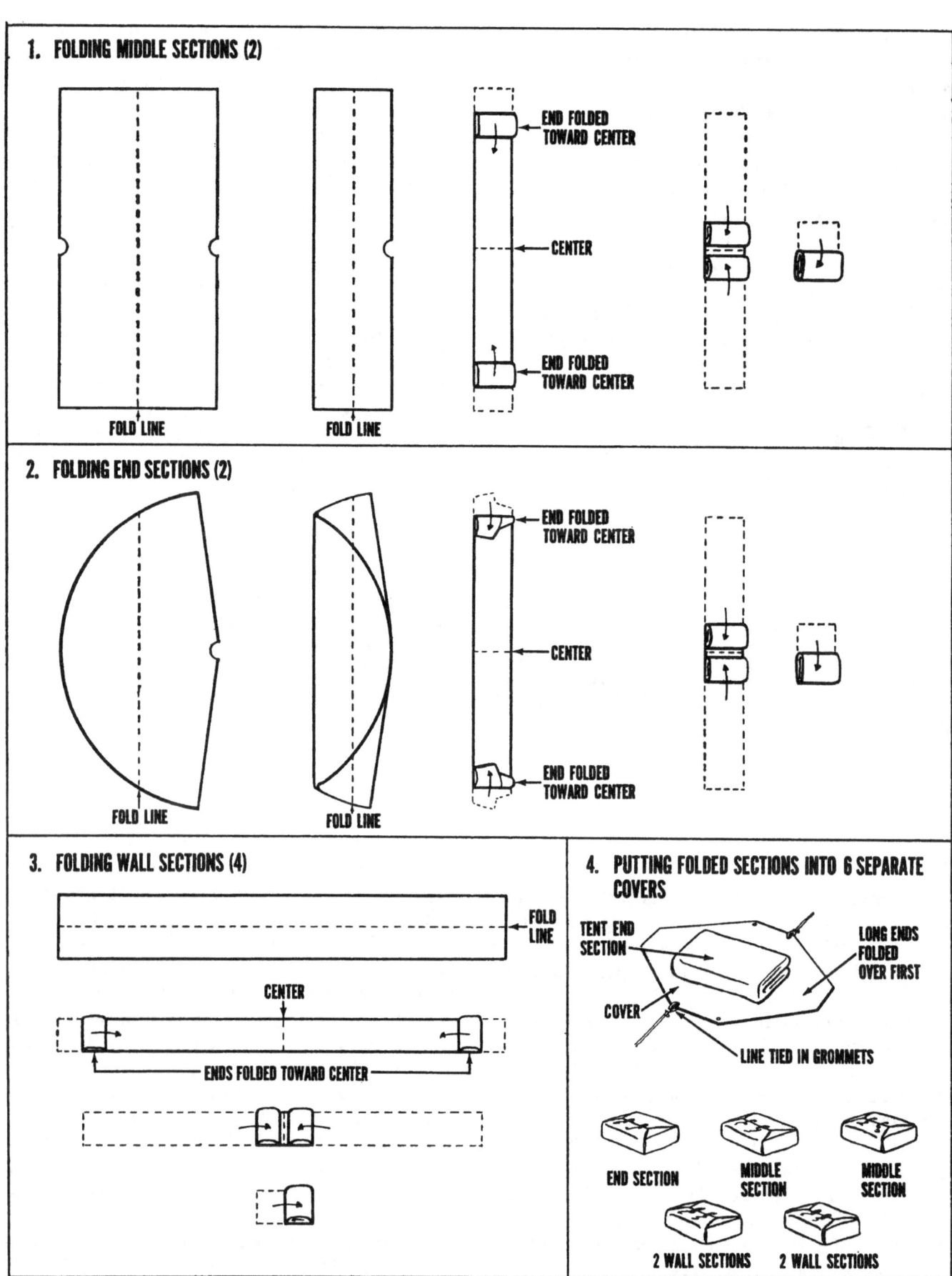

Figure 12. Steps in folding tent, assembly, M-1942, FWWMR, OD.

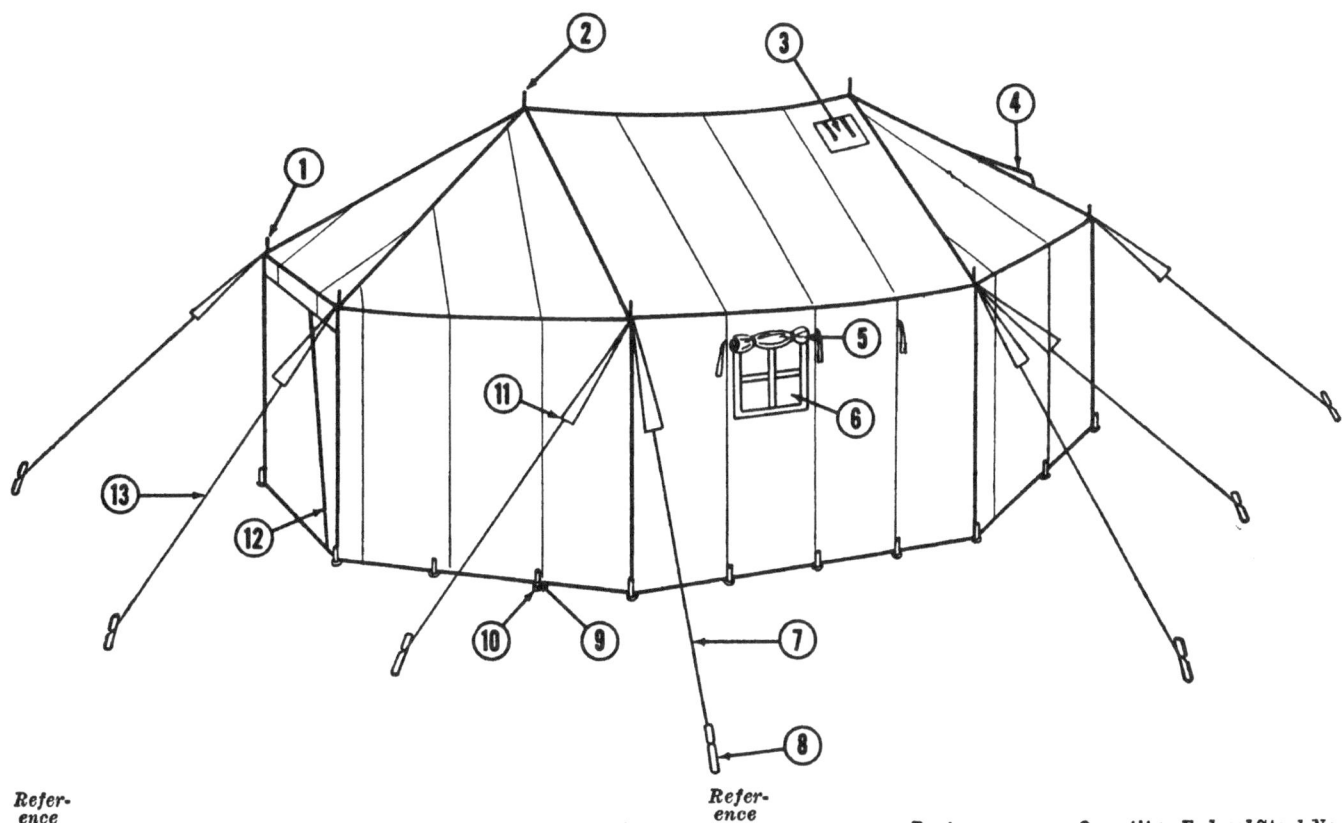

Reference No.	Part	Quantity	Federal Stock No.
1	Pole, tent, upright, solid, 5' 8" (eave pole)	8	8340-188-8405
2	Pole, tent, upright, solid, 9' (center pole)	2	8340-188-8410
3	Stovepipe opening	1	
4	Ventilator	1	
5	Window flap	3	
6	Window, w/o screening	2	
7	Line, tent, manila, one end sewed, one end w/eye, 5/16" dia., 11' 6" long (corner line)	8	8340-252-2268
8	Pin, tent, wood, 24"	12	8340-261-9751
9	Pin, tent, wood, 16"	20	8340-261-9750
10	Line, tent, cotton, footstop, 1/4" dia., 19" long (footstop)	24	8340-252-2299
11	Slip, tent line, steel	12	8340-223-8094
12	Vestibule door opening	1	
13	Line, tent, manila, one end sewed, one end w/eye, 5/16" dia., 13' long (door eave line)	4	8340-252-2271
	Line, tent, manila, one end sewed, 1/4" dia., 3' 4" long (jumper line)	8	8340-252-2291
	Line, tent, manila, one end sewed, 1/4" dia., 3' 4" long (door flap line)	2	8340-252-2291
	Line, tent, manila, one end sewed, 1/4" dia., 2' long (door tie line)	2	8340-252-2290
	Line, tent, manila, one end sewed, 1/4" dia., 6' long (jumper line)	2	8340-252-2286
	Window, w/screening	1	
	Tent screen, wall, command post, M-1945, FWWMR, OD	2	8340-262-3683
	Tent liner, command post, M-1945, FWWMR, OD	1	8340-262-3699
	Line, tent, cotton, 9' long, unfinished, 1/8" dia. (liner hoisting line)	2	8340-252-2301
	Cover, tent, FWWMR, command post, M-1945	1	8340-262-3654
	Line, tent, manila, one end sewed, one end w/eye, 5/16" dia., 13' long (cover tie line)	2	8340-252-2271

Figure 13. Tent, command post, M-1945, FWWMR, OD, complete with pins and poles, Federal Stock No. 8340-269-1370.

pole. When not in use, the opening can be protected by a canvas flap.

j. Liner. A liner is provided with the tent. The liner may be attached to the tent to insulate it against heat or cold. The liner, when attached, covers only the main part of the tent; it does not cover the vestibule.

k. Cover. The tent is provided with a cover for use when in storage or being transported.

17. Ground Plan

Before pitching the tent, study the ground plan carefully (fig. 14).

18. Pitching

The tent can be pitched by 5 men in approximately 20 minutes.

a. Pitching Tent.

(1) *Spreading tent on ground, driving eight*

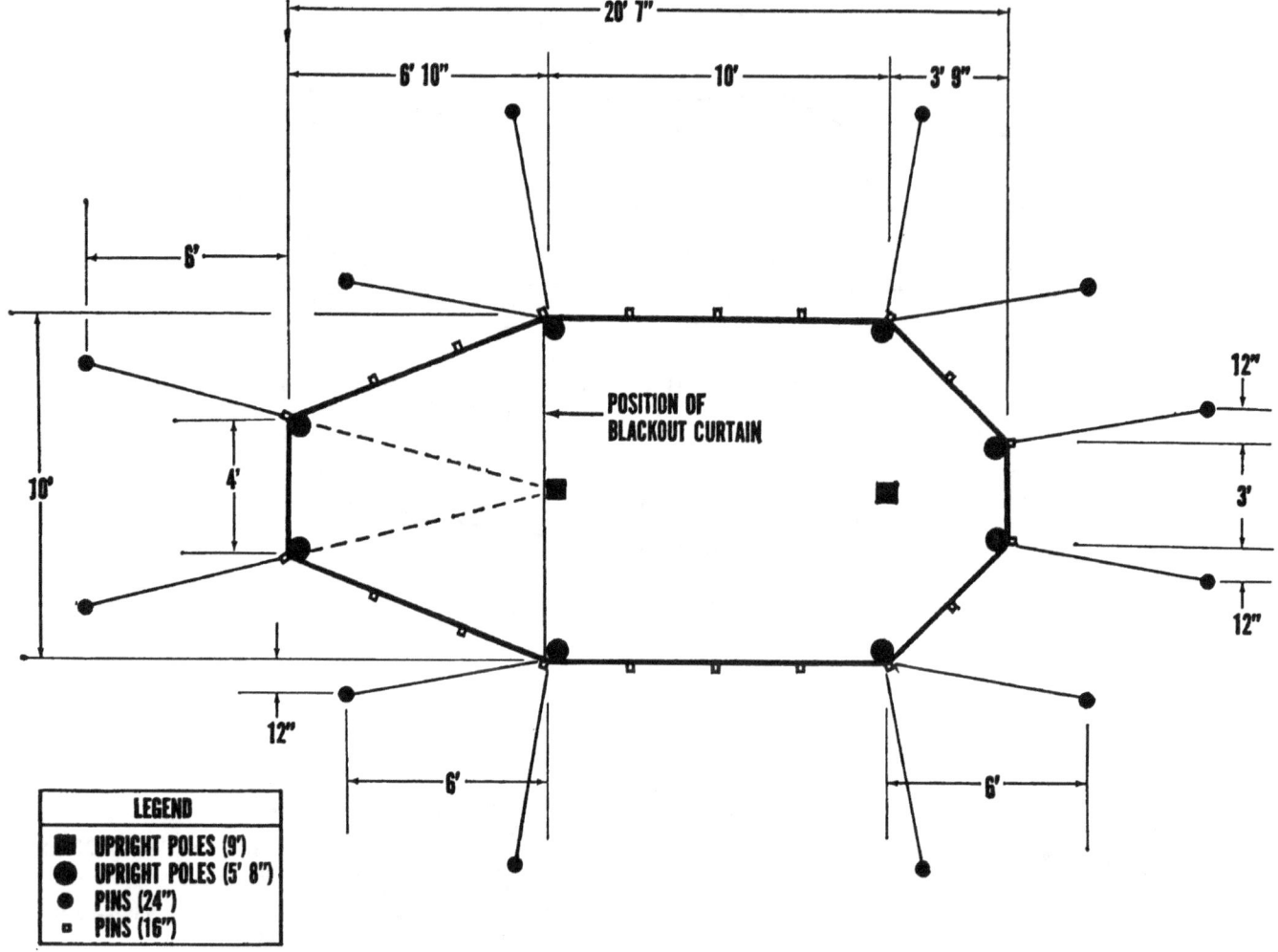

Figure 14. Ground plan of tent, command post, M-1945, FWWMR, OD.

16-inch pins, and attaching corner footstops to pins (1, fig. 15).

(a) Spread tent on ground with eave corners matching bottom corners.

(b) Drive a 16-inch pin at each of the 8 tent corners. Attach a footstop to each of the 4 corner pins at front and rear of tent. At each of the 4 side corners, attach 2 footstops to 1 pin.

(2) *Driving 24-inch pins and attaching guy lines* (2, fig. 15).

(a) Drive the twelve 24-inch pins according to ground plan.

(b) Attach guy lines loosely to long pins.

(3) *Removing corner footstops, inserting eave poles through grommets, and tightening guy lines* (3, fig. 15).

(a) Remove corner footstops from 16-inch pins.

(b) Insert eave poles through eave grommets.

(c) Tighten guy lines until poles are vertical.

(4) *Raising tent ridge by inserting center poles* (4, fig. 15). Raise tent ridge by inserting spindle of a 9-foot pole through hole in metal plate and grommet at ridge at front of tent, and spindle of the other 9-foot pole through hole in metal plate and grommet in other end of ridge.

(5) *Securing jumper lines to center poles and to eave poles, attaching footstops to short pins, and tightening guy lines* (5, fig. 15).

(a) Secure jumper lines to center poles and to eave poles with 2 half hitches.

(b) Reattach the 12 corner footstops to the 8 corner 16-inch pins.

(c) Drive the remaining twelve 16-inch pins and attach footstops to them.

(d) Tighten all guy lines.

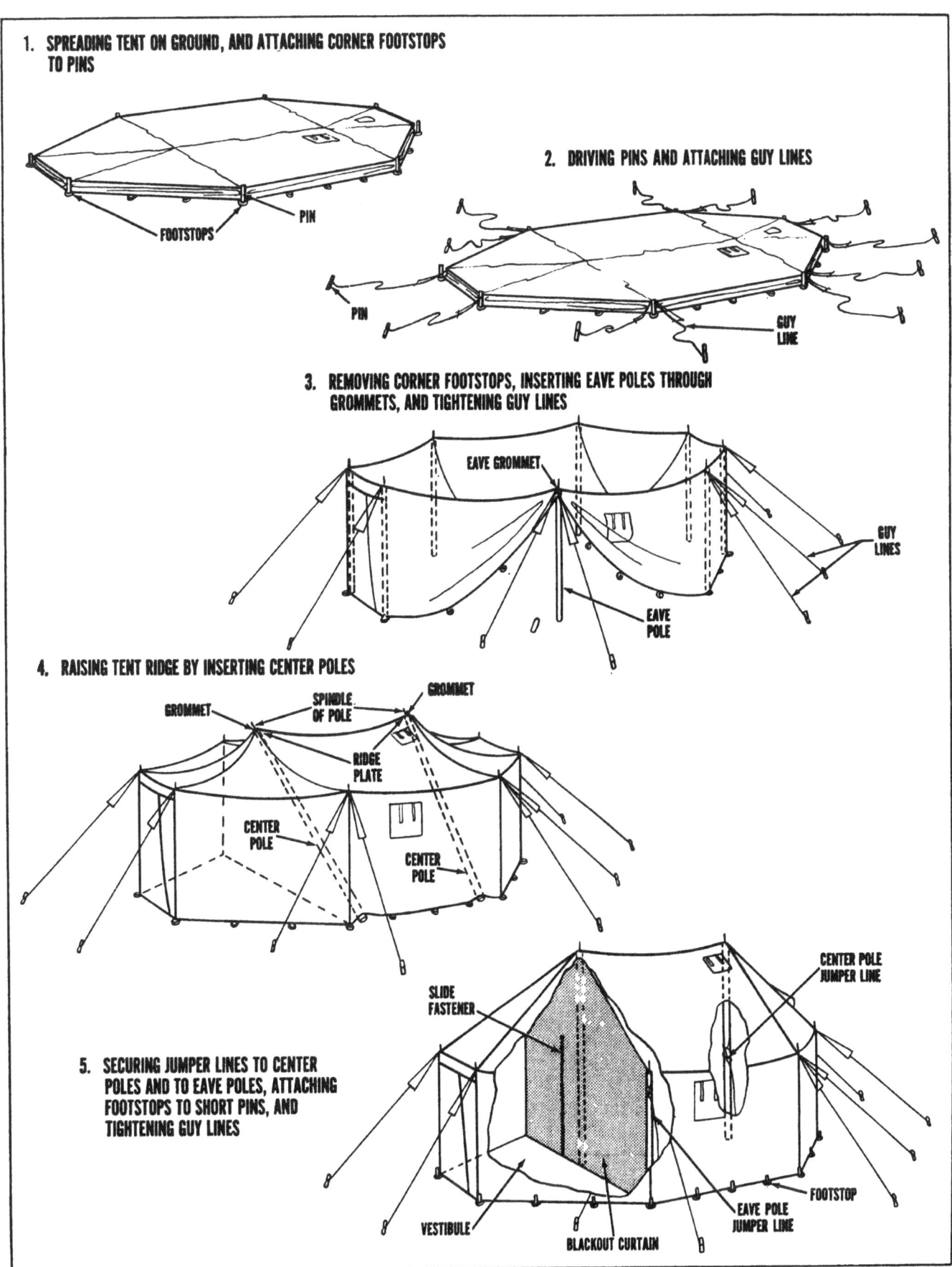

Figure 15. Steps in pitching tent, command post, M-1945, FWWMR, OD.

b. Attaching Liner to Tent.

(1) *Unrolling liner with stovepipe and window openings matching those of tent (1, fig. 16).* Unroll liner so that stovepipe and window openings match those of tent.

(2) *Placing rear center pole through hole in liner and running hoisting lines through bull's-eyes at tent ridge (2, fig. 16).*

(a) Raise butt end of rear center pole and place it through hole in liner.

(b) Run liner hoisting lines up poles through bull's-eyes at tent ridge.

(3) *Hoisting liner up to top of tent and se-*

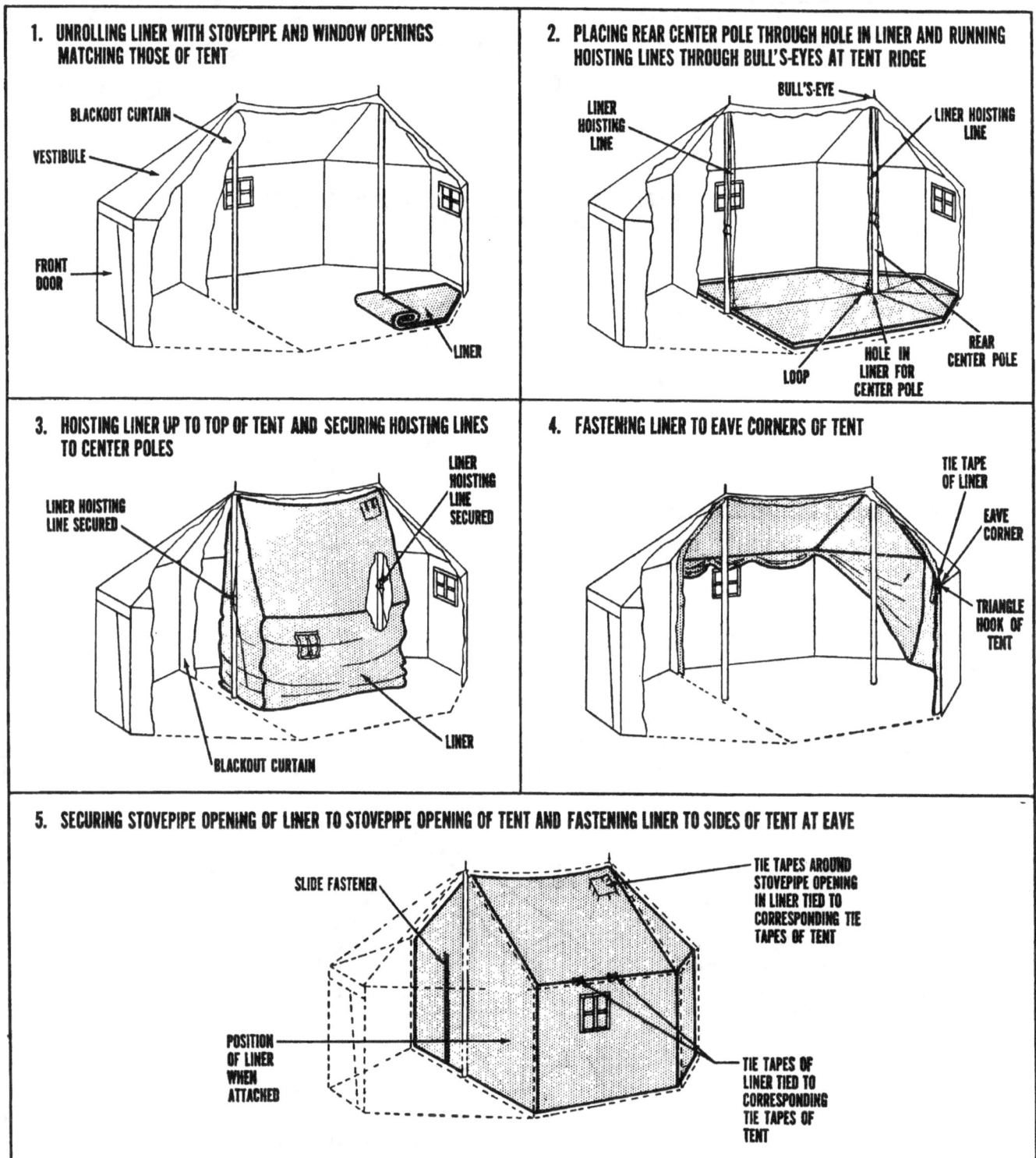

Figure 16. Steps in attaching liner to tent, command post, M-1945, FWWMR, OD.

curing hoisting lines to center poles (3, fig. 16). Hoist liner up to top of tent and secure lines to center poles.

(4) *Fastening liner to eave corners of tent* (4, fig. 16). Fasten liner to eave corners of tent by tying tie tapes of liner to triangular hooks of tent.

(5) *Securing stovepipe opening of liner to stovepipe opening of tent and fastening liner to sides of tent at eave* (5, fig. 16).

  (a) Tie tie tapes at stovepipe opening of liner to corresponding tie tapes at stovepipe opening of tent.

  (b) Tie tie tapes at sides of liner at eave to corresponding tie tapes at sides of tent at eave.

c. *Attaching Screens to Side Walls* (fig. 17). Remove footstops from side walls; open slide fasteners at corners; roll up side walls and liner of tent and tie them with tie tapes near eave reinforcement. Then place a screen between lugs at each side of tent, alining grommets on screen with grommets on lugs. Run the 13-foot rope, attached to a top corner of screen, through alined grommets of screen and lugs at the top, securing with a knot at the last set of alined grommets. In the same manner, run the 8-foot rope, attached to each side of screen, downward through alined grommets of screen and lugs, securing with a knot at the last set of alined grommets. Fold screens at bottom so sod cloths are on ground inside tent. Fasten footstops that were removed from side walls to grommets at bottoms of screens, and attach footstops to the 16-inch pins.

19. Striking

a. Remove screens, and lower side walls of tent.
b. Loosen liner hoisting lines, and untie tapes fastening liner to tent.
c. Remove liner.
d. Remove all footstops from 16-inch pins.
e. Loosen all guy lines and remove center poles.
f. Remove all 5-foot 8-inch eave poles.
g. Remove all guy lines from 24-inch pins.
h. Remove all pins.

20. Folding

a. *Folding Tent at Ridge, With Sides Together Flat* (1, fig. 18). Fold tent at ridge, with tent laid out flat one side on top of the other, sod cloth and vestibule door flaps extended, and blackout curtain folded neatly one half on top of the other half.

b. *Folding Over Door Flaps and Rear of Tent.* (2, fig. 18).

  (1) Fold door flaps over on top of vestibule.
  (2) Fold rear of tent over body of tent, the fold line extending from rear ridge plate down along rear body slide fasteners.

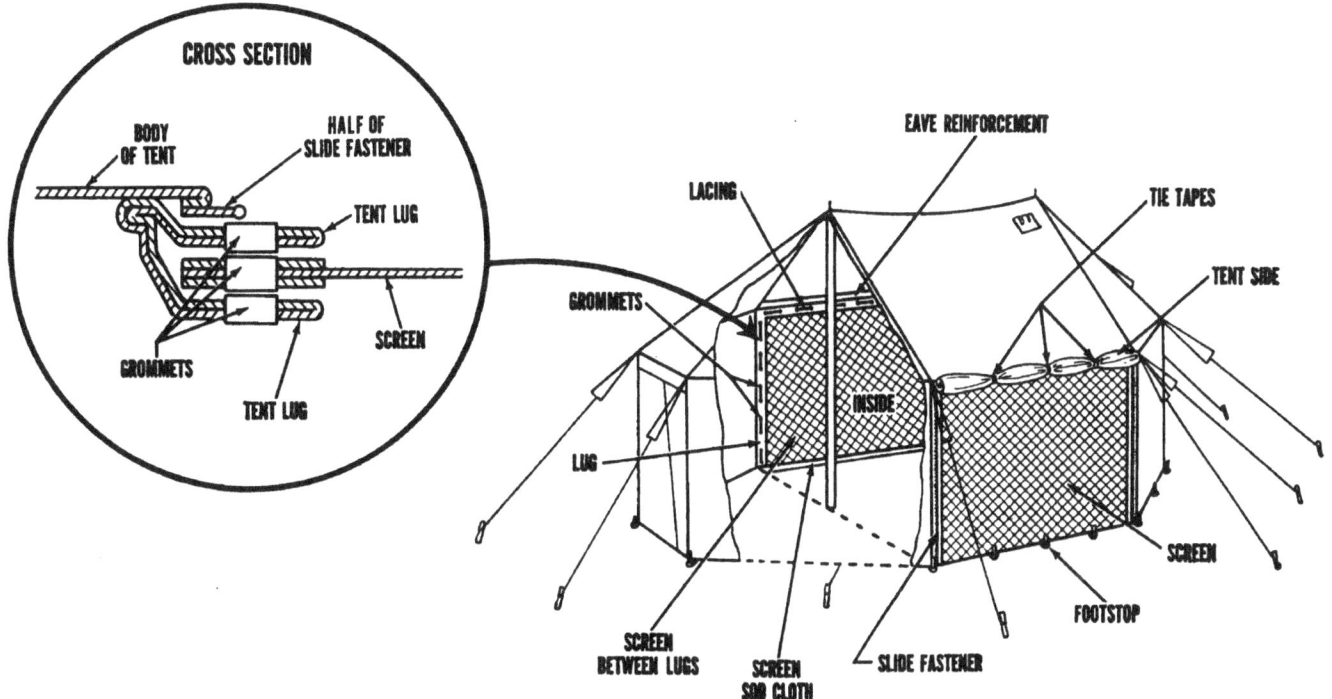

Figure 17. Attaching screens to side walls of tent, command post, M-1945, FWWMR, OD.

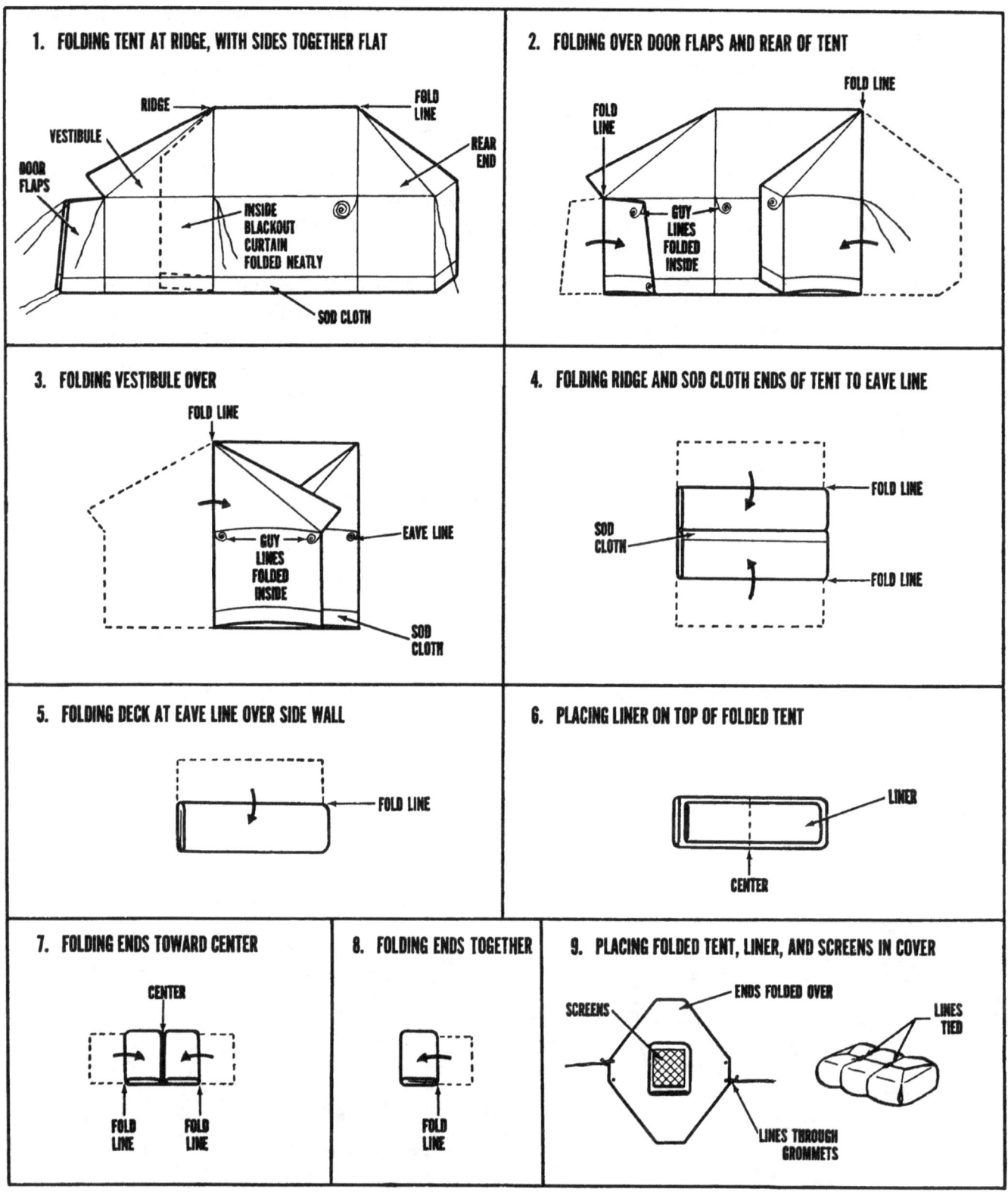

Figure 18. Steps in folding tent, command post, M-1945, FWWMR, OD.

(3) Fold guy lines inside folded tent.

c. *Folding Vestibule Over* (3, fig. 18). Fold vestibule over body of tent, the fold line extending from front ridge plate down along front body slide fasteners.

d. *Folding Ridge and Sod Cloth Ends of Tent to Eave Line* (4, fig. 18). Fold both the ridge end of tent and the sod cloth end of tent to the eave line.

e. *Folding Deck at Eave Line Over Side Wall*

(5, fig. 18). Fold deck of tent at eave line over side wall. Place exposed guy lines on folded tent.

f. Placing Liner on Top of Folded Tent (6, fig. 18). Place liner, folded in the same way as the tent, on top of tent.

g. Folding Ends Toward Center (7, fig. 18). Fold ends of folded tent and liner toward center.

h. Folding Ends Together (8, fig. 18). Fold the two ends together.

i. Placing Folded Tent, Liner, and Screens in Cover (9, fig. 18).
 (1) Place folded tent and liner in cover.
 (2) Place screens, each folded in fourths, on top of folded tent and liner.
 (3) Close cover, and tie with the 2 tie lines.

Section IV. GENERAL PURPOSE TENT, LARGE

21. Use

The tent, general purpose, large, FWWMR, OD, complete with pins and poles (fig. 19), is used to fulfill requirements for large tentage for storage purposes and shelter, as for a small bakery or a hospital ward.

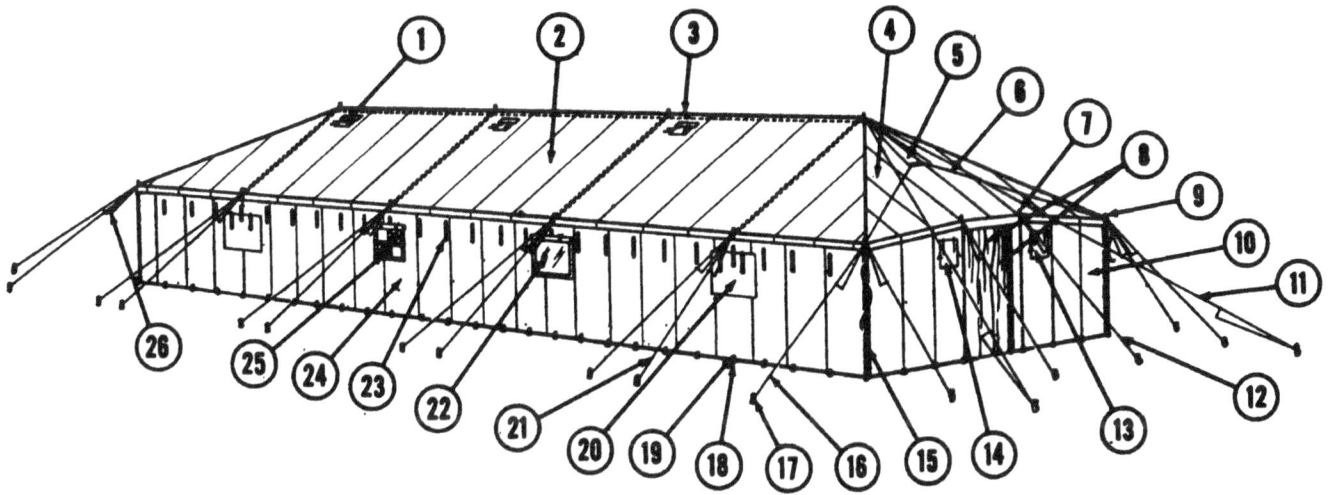

Reference No.	Part	Quantity	Federal Stock No.
1	Stovepipe opening	3	
2	Side roof		
3	Pole, tent, upright, jointed, metal sleeve, 12' 3" long (center pole)	4	8340-188-8411
4	End roof		
5	Ventilator	2	
6	Line, tent, manila, one end sewed, ¼" dia., 14' long (ventilator flap line)	4	8340-252-2280
7	Pole, tent, upright, solid, 6' 2" long (door pole)	4	8340-188-8406
8	Door curtain	4	
9	Pole, tent, upright, solid, 5' 8" long (eave pole)	12	8340-188-8405
10	End wall		
11	Line, tent, manila, both ends sewed, 5/16" dia., 50' long (ridge guy line)	2	8340-252-2293
12	Line, tent, manila, one end sewed, one end with eye, 5/16" dia., 13' long (eave guy line, door)	4	8340-252-2271
13	Care and maintenance instructions flap	1	
14	Erection instructions flap	1	
15	Fastener, slide, interlocking nonseparating, open top stop, closed bottom stop, medium heavy duty	4	Unassigned
16	Line, tent, manila, one end sewed, one end with eye, 5/16" dia., 13' long (eave guy line, corner)	8	8340-252-2271
17	Pin, tent, wood, 24", or pin, tent, steel, 12" (for cold climate use)	32 / 32	8340-261-9751 / 8340-261-9748
18	Line, tent, cotton, footstop, ¼" dia., 19" long	72	8340-252-2299
19	Pin, tent, wood, 16", or pin, tent, aluminum, 9" (for cold climate use)	68 / 68	8340-261-9750 / 8340-261-9749
20	Window blackout flap	8	
21	Line, tent, manila, one end sewed, one end with eye, 5/16" dia., 13' long (eave guy line, side)	16	8340-252-2271
22	Windowpane	8	
23	Line, wall, 13½" (tie tape)		
24	Side wall		
25	Window screen	8	
26	Slip, tent line, steel	32	8340-223-8094
	Line, tent, manila, one end sewed, ¼" dia., 3' 4" long (eave jumper line)	16	8340-252-2291
	Line, tent, manila, one end sewed, ¼" dia., 6' long (ridge jumper line)	4	8340-252-2286

Figure 19. Tent, general purpose, large, FWWMR, OD, complete with pins and poles, Federal Stock No. 8340-285-5599.

22. Description

The tent is a hip-roofed, square-end rectangular tent.

a. Dimensions. The tent is 18 feet wide, 52 feet long, and 12 feet high. The wall height is 5 feet 6 inches.

b. Weight and Cubage. The tent weighs 420 pounds; the liner, 155 pounds; and the pins and poles, 245 pounds. The tent, liner, pins, and poles have a cubage in storage of 69 cubic feet.

c. Floorspace. The floorspace of the tent is 936 square feet.

d. Material. The roof, side walls, and end walls are made of 12.29-ounce cotton duck, FWWMR. The whole tent is made in one piece. The canvas is suspended on a webbing framework, which carries the stress and supports the canvas. The walls are split at the 4 corners and can be fastened together with a slide fastener at each corner.

e. Doors. The tent has 2 door entrances, one at each end. Each door entrance is 6 feet high and 4 feet wide.

   (1) *Door curtains.* Two curtains, attached to each end near the door entrances, slide along a double wire cable at the eave to open or shut the door entrances.

   (2) *Door screens.* A screen is attached on the inside to each side of each door entrance. When in use, the door screens are pulled across the door entrances and secured in place by tying tie tapes at the top of the screens to metal rings at the eave above the door entrances. When not in use, the door screens are rolled to the side inside the tent and secured by tying the screens with the tie tapes at one side of the door.

f. Windows. There are 4 window assemblies on each side of the tent below the eave. Each window assembly consists of a plastic window screen, a vinyl plastic windowpane, and a canvas blackout flap. The window screen is attached to the side wall. The windowpane is attached at the top to the side wall and is secured at the bottom and the 2 sides by a slide fastener. The slide fastener may be unfastened and the windowpane rolled up and tied at the top with tie tapes. A blackout flap is attached at the top to the side wall. When the flap is in use, it is secured by tying tie tapes at the 2 sides; when not in use, it is rolled up and tied at the top with the tapes.

g. Ventilation.

   (1) The tent is ventilated by 2 ventilators, one at the top of each end section near the ridge. The openings are protected by canvas flaps.

   (2) When stoves are not being used, the stovepipe openings may also be used as ventilators.

   (3) Additional ventilation can be obtained by rolling up the window blackout flaps and the windowpanes and tying them open with tie tapes.

   (4) The door curtains can be opened for more ventilation.

   (5) Additional ventilation can be obtained by rolling up the sides of the tent to the eaves and tying them with tie tapes.

h. Heating. The tent is heated by three M-1941 tent stoves. There are three stovepipe openings built in the top of the tent. Each opening is protected by a canvas flap.

i. Cover. The tent is provided with a cover for use when in storage or when being transported.

j. Liner. A liner (Federal Stock No. 8340-285-5033) is available as a separate item of issue. It provides insulation from the cold in winter and reduces radiation from the sun in summer. The liner weighs 155 pounds. The liner may be attached to the inside of the tent at the eaves by passing the eave suspension line on the liner through hardware eye or around webbing on inside of end roof of tent, and then running line through grommet in liner and securing to D-ring. The liner is attached at the ridge by tying ridge suspension lines at liner pole openings to hardware at upright poles and then tying ridge suspension lines at ridge of liner to D-rings which are located along tent ridge. The liner has 5.2-ounce cotton cloth side walls below the eaves and, in addition, has screening side walls made of plastic. The fabric side walls may be rolled up to the eaves and secured by tie tapes and thus permit the use of the screening alone. The screening provides protection from insects and permits the liner to be used in hot as well as cold weather. There are 2 built-in ventilator screens corresponding in location to the 2 ventilators in the tent. There are 4 vinyl plastic windows on each fabric side wall corresponding in location to the windows in the tent. There are 3 stovepipe openings in the liner corresponding in location to the stovepipe openings in the tent.

23. Ground Plan

Before pitching the tent, study the ground plan carefully (fig. 20).

Note. The center poles will be placed 2 feet off center to create an unobstructed aisle running the length of the tent.

24. Pitching

a. Pitching Tent. Six men can pitch the tent in approximately 1¼ hours.

(1) *Spreading tent out, closing slide fasteners and doors, driving corner pins, and attaching corner footstop to pins* (1, fig. 21).

   (a) Spread tent out on ground into position, with corners square.

   (b) Close slide fasteners at corners.

   (c) Close doors by sliding door curtains across door entrances and fastening loops to wooden toggles.

   (d) Drive a 16-inch wood pin, or in cold climate a 9-inch aluminum pin, at each corner, and attach corner footstops to pins.

(2) *Driving eave-line pins, attaching eave lines, and placing eave and door poles in position for erection* (2, fig. 21).

   (a) Drive the 24-inch wood pins, or in cold climate the 12-inch steel pins, according to ground plan, using 5-foot 8-inch poles to measure distance out from tent.

   (b) Attach side, corner, and door eave lines loosely to pins.

   (c) Place the 5-foot 8-inch side and corner eave poles and the 6-foot 2-inch door poles in position for erection.

(3) *Inserting eave poles through handworked rings and raising tent walls* (3, fig. 21).

   (a) Insert spindles of the 5-foot 8-inch eave poles through handworked ring at sides and corners of tent.

   (b) Insert spindles of the 6-foot 2-inch door poles through handworked ring at front and rear doors.

   (c) Raise tent walls by raising side, corner, and door eave poles to an upright position.

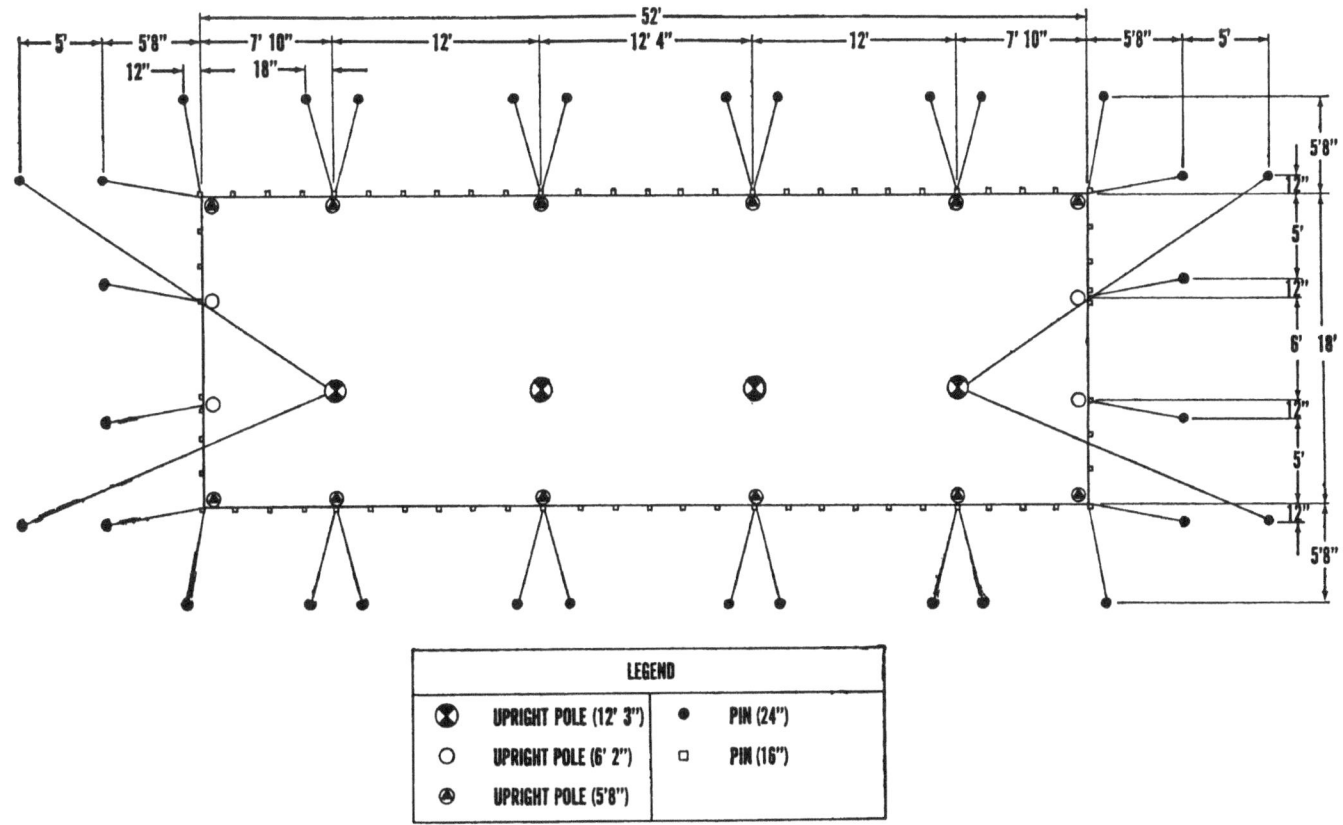

Figure 20. Ground plan of tent, general purpose, large, FWWMR, OD, complete.

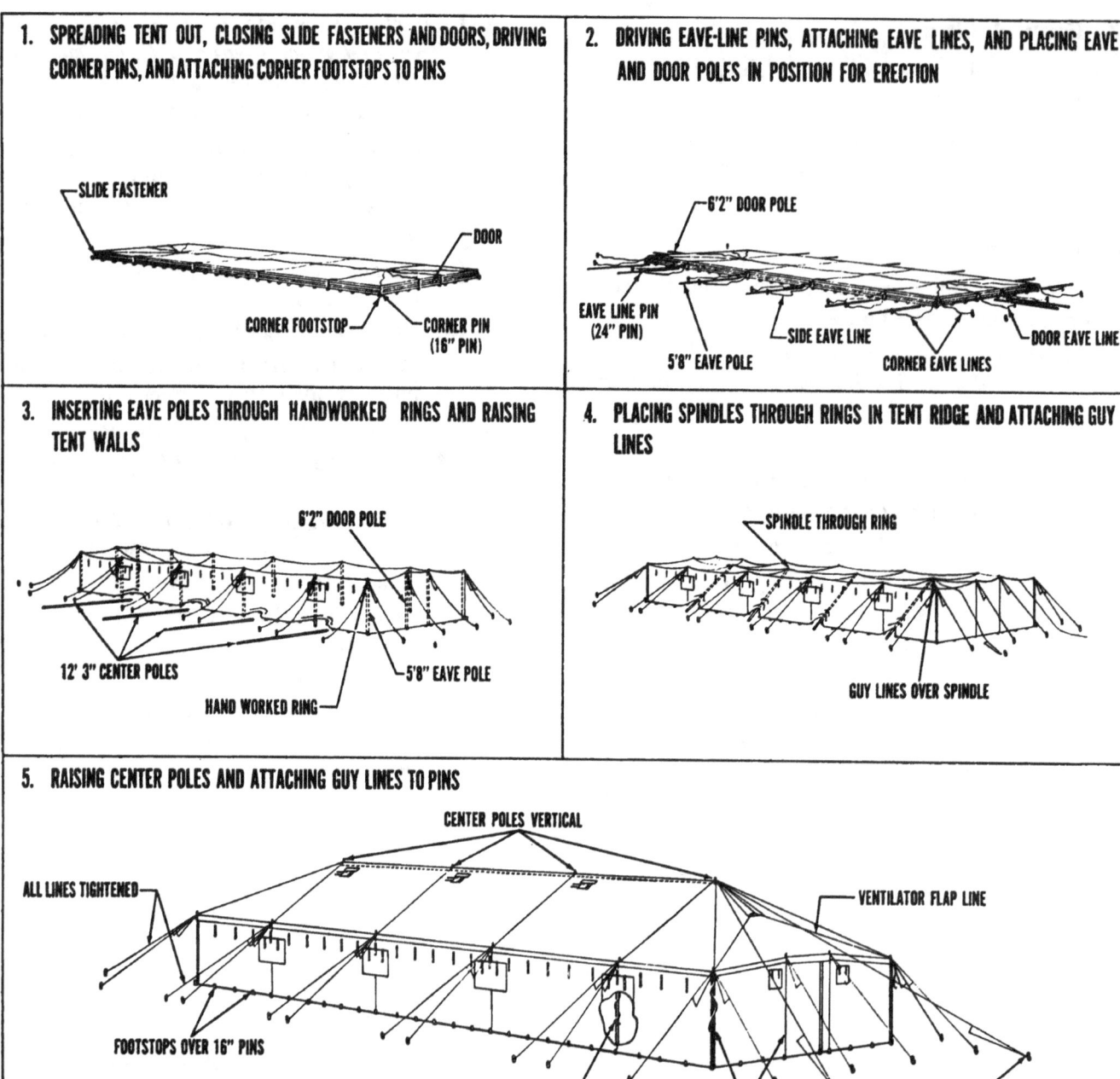

Figure 21. Steps in pitching tent, general purpose, large, FWWMR, OD.

(d) Tighten eave lines just enough to hold poles up.

(4) *Placing spindles through rings in tent ridge and attaching guy lines* (4, fig. 21).

   (a) Assemble the 12-foot 3-inch center poles, and insert spindles of poles in handworked rings in ridge of tent.

   (b) Attach guy lines to spindles of center poles at each end of tent.

(5) *Raising center poles and attaching guy lines to pins* (5, fig. 21).

   (a) Raise the 4 center poles to a vertical position.

   (b) Attach guy lines to pins and tighten.

   (c) Drive remaining 16-inch wood pins, or in cold climate the 9-inch aluminum pins, and attach footstops to pins.

   (d) Tie jumper lines to side and corner eave poles, door poles, and center poles.

   (e) Adjust ventilator flap lines and tie them to spindles of corner eave poles.

   (f) Straighten all poles.

(g) Tighten all lines and adjust tent slips until tent is smooth.

(h) Tie tie tapes at inside corners of tent around corner eave poles.

b. *Attaching Liner to Tent.*

(1) *Loosening tent lines* (1, fig. 22). Loosen slightly all tent lines by adjusting tent slips.

(2) *Unrolling liner inside tent* (2, fig. 22). Unroll liner inside tent on one side of center poles, with stovepipe openings of liner matching those of tent.

(3) *Placing liner around center poles and spreading liner* (3, fig. 22). Lift each center pole and pull liner under base of pole until hole in liner is directly under pole; then set pole down and spread liner into position.

(4) *Pulling up and securing liner to ridge* (4, fig. 22). Tie ridge suspension lines at liner pole openings to hardware at upright poles; then tie rope suspension lines at ridge of liner to **D**-rings located along tent ridge.

(5) *Fastening liner at doors* (5, fig. 22). Fasten liner to tent at doors by passing eave suspension line on liner to hardware eye or around webbing on each side of end roof of tent; then run line through No. 4 grommet in liner and secure to **D**-ring. Tie tie tapes at sides of door opening to door eave poles.

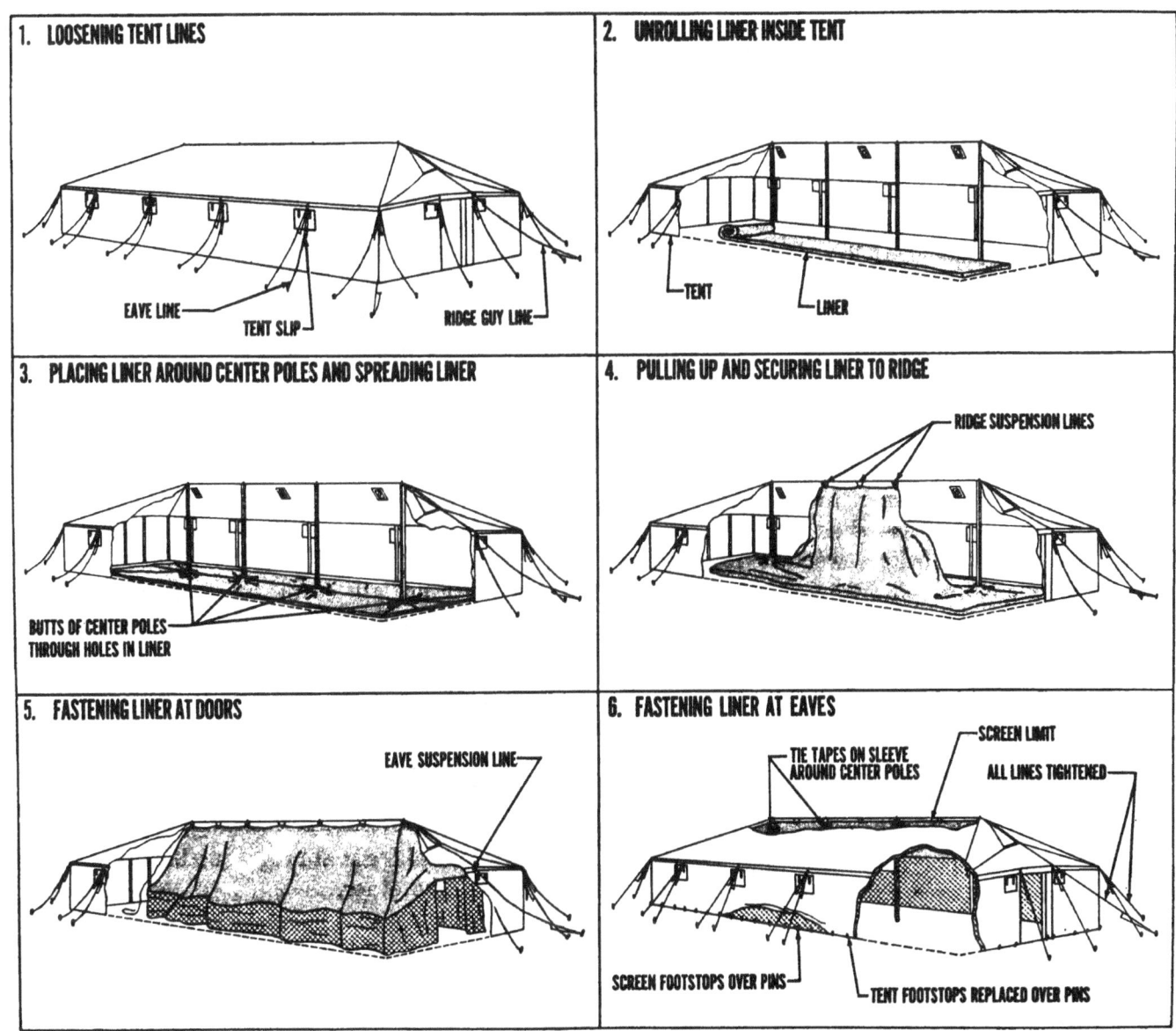

Figure 22. Steps in attaching liner to tent, general purpose, large, FWWMR, OD.

(6) *Fastening liner at eaves* (6, fig. 22). Fasten liner to tent along eave with eave suspension lines. Fasten sleeves to poles with tie tapes. Finish securing all tie tapes, and tighten all tent guy lines after securing footstops in side wall screen to tent footstop pins.

25. Striking

Six men can strike the tent in approximately 50 minutes.

  a. Removing Liner.

    (1) Unfasten footstops, both of tent and of liner screen, from the 16-inch wood or the 9-inch aluminum pins.

    (2) Untie tie tapes at corners. Untie tie tapes at door entrances from door eave poles.

    (3) Unfasten eave suspension lines from D-rings, pull suspension lines through grommets to outside of liner, and unfasten suspension lines from webbing of tent or the hardware.

    (4) Untie tie tapes of sleeve from around center poles at top.

    (5) Untie suspension lines from hardware and from the 3 D-rings at ridge, and allow liner to drop to ground.

    (6) Lift center poles slightly and remove liner from the poles.

    (7) Fold liner as described in paragraph 26.

  b. Striking Tent.

    (1) Untie tie tapes at inside corners of tent from around corner eave poles.

    (2) Close the 4 corner slide fasteners. Close doors and fasten wooden toggles to toggle loops.

    (3) Remove all but the 4 corner 16-inch wood, or 9-inch aluminum, footstop pins.

    (4) Unfasten from the 24-inch wood pins, or the 12-inch steel pins, door eave lines and all other eave lines except those at corners. Remove the 24- or 12-inch pins from which door and side eave lines were unfastened.

    (5) Remove door eave poles and all other eave poles except those at corners.

    (6) Remove guy lines from the 24-inch wood pins, or the 12-inch steel pins. Lower center poles gently to ground. Remove the 24-inch wood pins, or the 12-inch steel pins.

    (7) Unfasten the 8 corner eave lines from the 24-inch wood pins, or the 12-inch steel pins, and remove corner eave poles and tent pins.

26. Folding

  a. Folding Liner (fig. 23).

    (1) Lay liner out as flat as possible with eave suspension lines on top, and side and end wall screens and side and end walls folded under toward center (A, fig. 23).

    (2) Fold ridge peak of triangular end pieces toward eave line (B).

    (3) Starting at one end, make a 3½-foot fold toward middle of tent (C).

    (4) Measuring from fold (outer edge) of previous fold, make a second fold of 6 feet toward middle of tent (C).

    (5) Make two additional 6-foot folds toward middle of tent (C).

    (6) Starting at opposite end of tent make the same folds until the last fold is placed on top of the previously folded section. Dimensions of folded liner at this point are approximately 6 by 18 feet (D).

    (7) Fold liner in thirds along the 18-foot length, placing the last fold over the first fold. At this point, dimensions are approximately 6 by 6 feet (D). Care should be taken that folds do not come at windows.

    (8) Fold liner in half along each 6-foot edge (D).

    (9) Place the completely folded liner in its cover (D).

    (10) Fold all cover ends or flaps neatly within package, and close cover securely.

  b. Folding Tent.

    (1) *Preliminary procedures* (1, fig. 24).

      (*a*) Spread tent out flat with inside facing up.

      (*b*) Close and secure all doors, window assemblies, and stovepipe openings.

      (*c*) Fold tent at ridge, one-half folded over the other.

      (*d*) Coil eave lines toward center.

    (2) *Folding end walls over toward center* (2, fig. 24). Fold end walls over on deck of tent and corner eave guy lines toward center of tent.

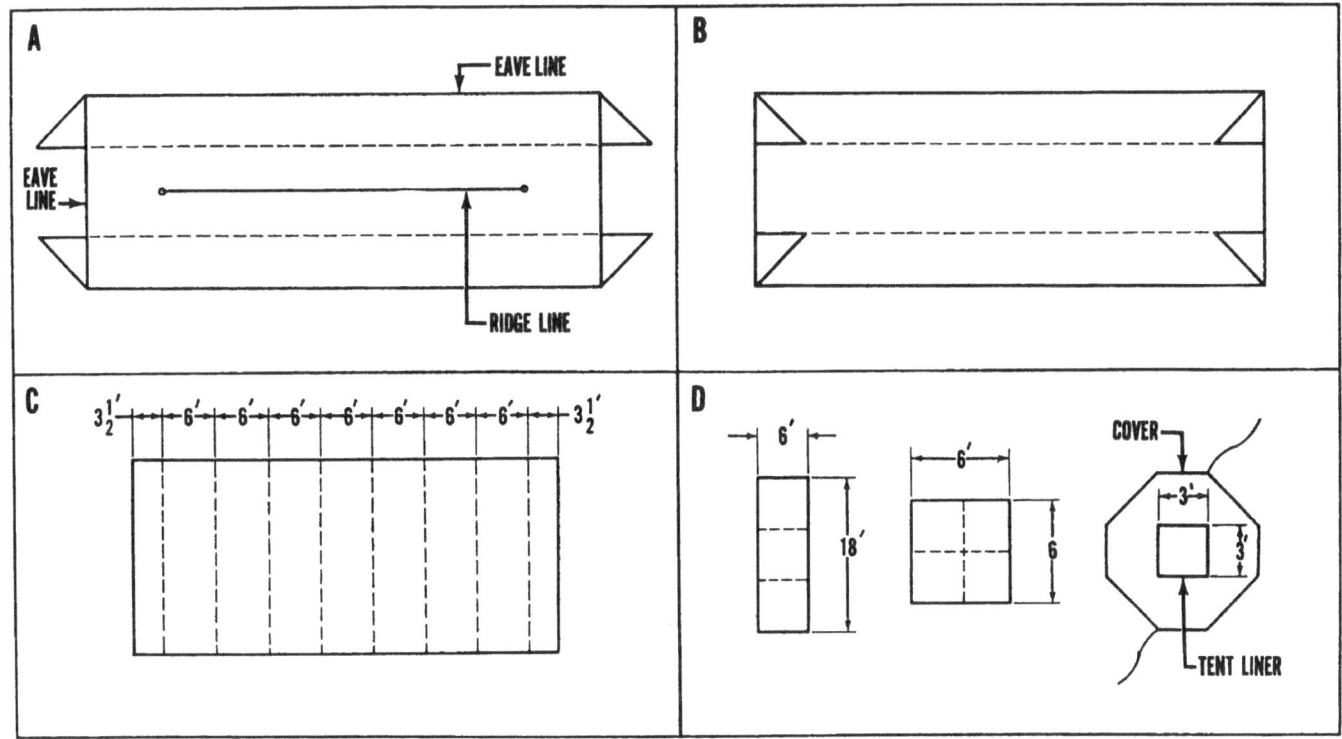

Figure 23. Steps in folding liner of tent, general purpose, large, FWWMR, OD.

(3) *Folding roof over so ridge is at eave* (3, fig. 24). Bring ridge of tent over and place it on side wall at eave.

(4) *Folding roof over again to eave and folding side walls to eave in two folds* (4, fig. 24).

  (a) Bring fold line, formed by placing ridge on side wall, over and place it on top of ridge.

  (b) Fold side walls to eave in 2 folds, making sure the second fold line is directly below windows.

(5) *Folding bulk of tent over side wall* (5, fig. 24). Fold bulk of tent (folded roof) over side walls, exposing eave. Then coil remaining eave lines on top of folded tent.

(6) *Folding each end toward center and inserting guy lines* (6, fig. 24). Fold ends of tent toward center, forming two folded sections approximately 1 foot apart at center. Care should be taken that folds do not come at windows. Then neatly coil ridge guy lines, tie in middle with suitable cord or twine, and place upon either folded section.

(7) *Folding end over end and placing in cover* (7, fig. 24). Fold end over end and place in cover. Close cover securely.

Section V. GENERAL PURPOSE TENT, MEDIUM

27. Use

The tent, general purpose, medium, FWWMR, OD, complete with pins and poles (fig. 25), is used principally for the quartering of men. It can shelter 12 men when stoves are installed or 16 men when stoves are not installed. It may also be used as a divisional or regimental command post, first-aid station, mess tent, or for any other authorized purpose.

28. Description

The tent is a hip-roofed, square-end rectangular tent.

a. Dimensions. The tent is 16 feet wide, 33 feet long, and 10 feet high. The wall height is 5 feet 6 inches, giving a pitch of 4 feet 6 inches.

b. Weight and Cubage. The tent weighs 255 pounds and the pins and poles 200 pounds. The tent has a cubage in storage of 12.7 cubic feet,

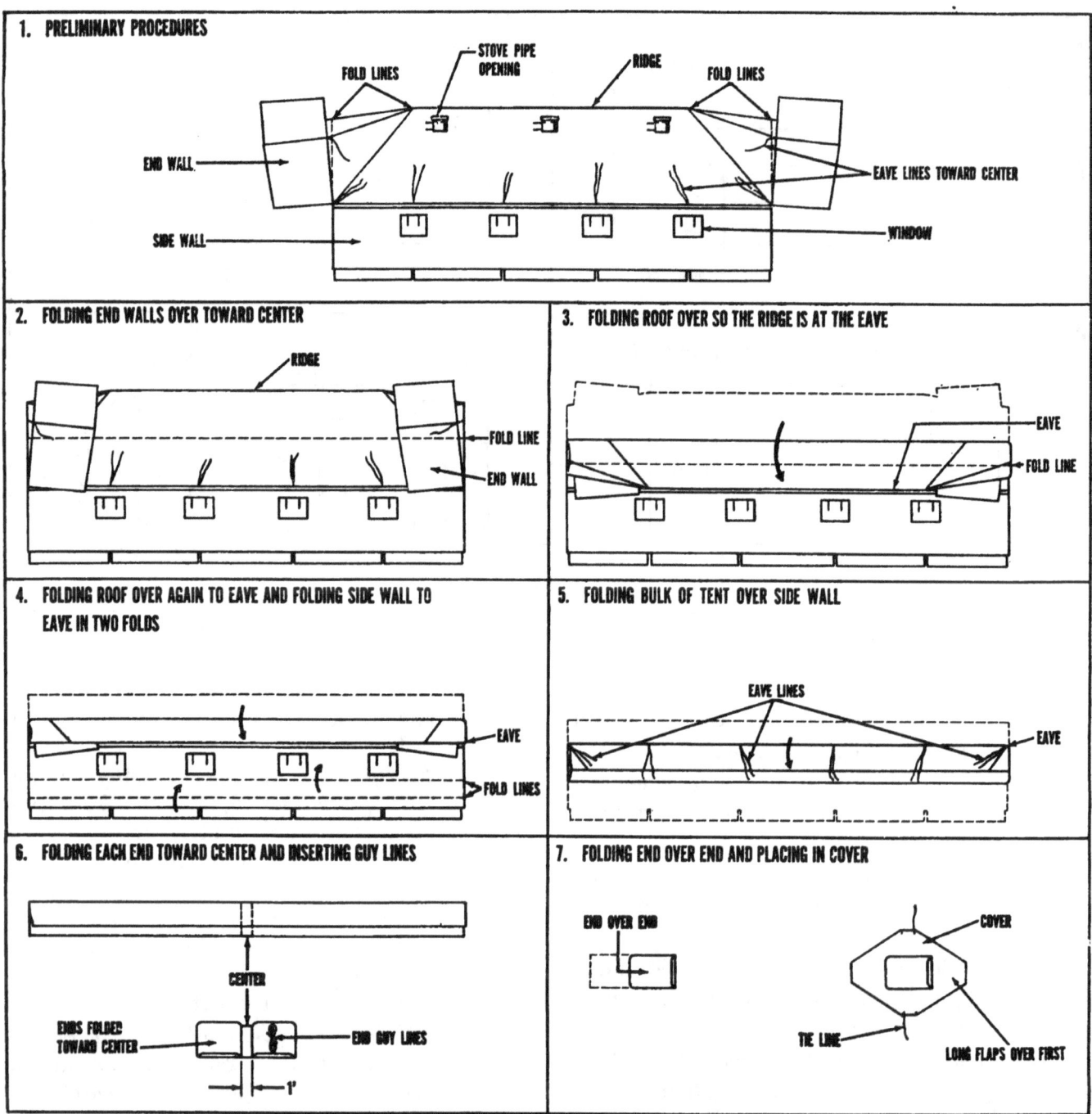

Figure 24. Steps in folding tent, general purpose, large, FWWMR, OD.

and the pins and poles have a cubage of 6.3 cubic feet.

c. Floorspace. The floorspace of the tent is 528 square feet.

d. Materials. The roof, side walls, and end walls are made of 12.29-ounce duck, FWWMR. The whole tent is made in one piece. The canvas is suspended on a webbing framework which carries the stress and supports the canvas. The walls are split at the 4 corners and can be fastened together with a slide fastener at each corner.

e. Doors. The tent has 2 door entrances, one at each end. Each door entrance is 6 feet high and 4 feet wide.

    (1) *Door curtains.* Two curtains, attached to each end near the door entrances, slide along a double wire cable at the eave to open or shut the door entrances.

    (2) *Door screens.* A screen is attached on

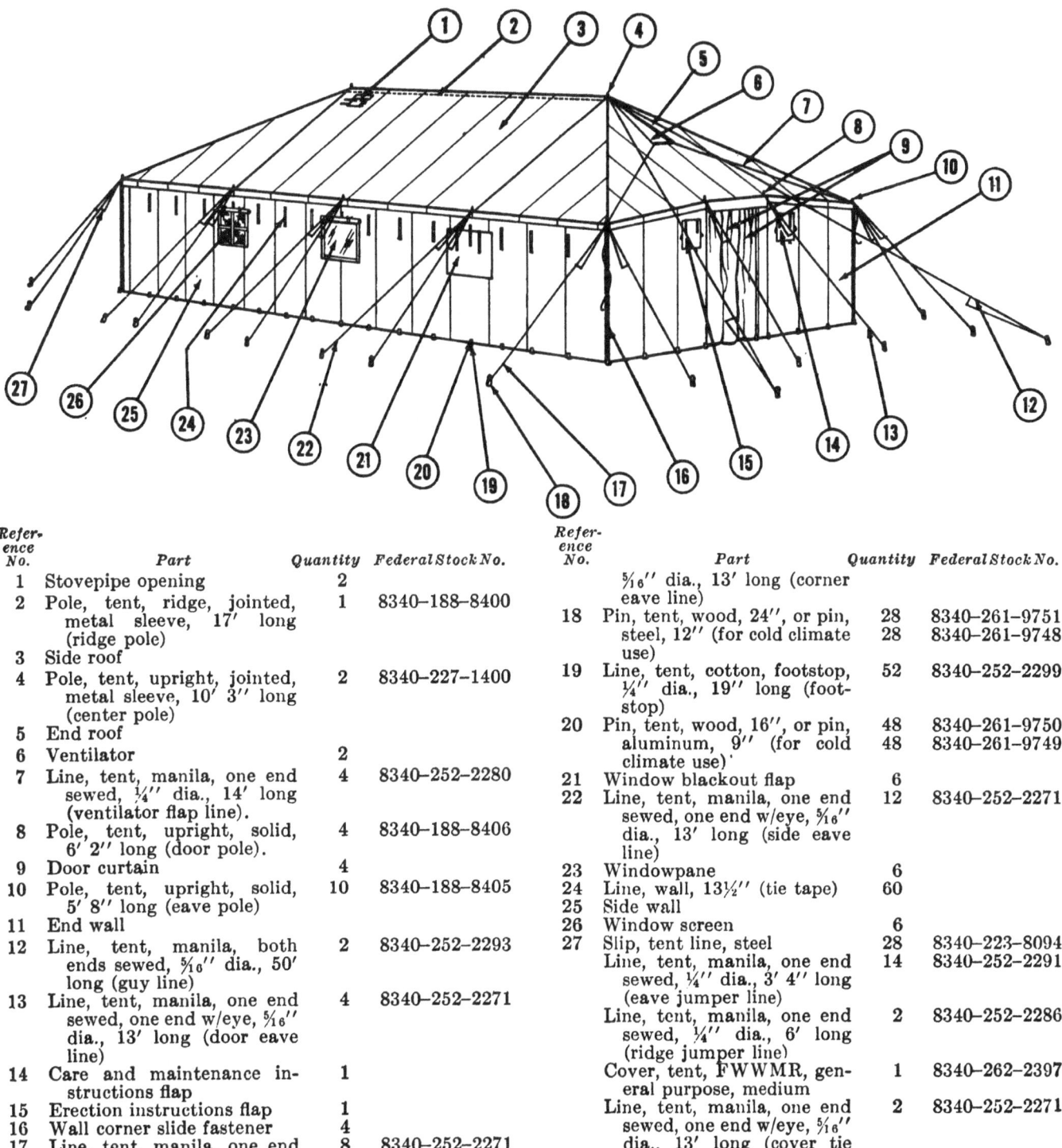

Reference No.	Part	Quantity	Federal Stock No.
1	Stovepipe opening	2	
2	Pole, tent, ridge, jointed, metal sleeve, 17′ long (ridge pole)	1	8340-188-8400
3	Side roof		
4	Pole, tent, upright, jointed, metal sleeve, 10′ 3″ long (center pole)	2	8340-227-1400
5	End roof		
6	Ventilator	2	
7	Line, tent, manila, one end sewed, ¼″ dia., 14′ long (ventilator flap line)	4	8340-252-2280
8	Pole, tent, upright, solid, 6′ 2″ long (door pole)	4	8340-188-8406
9	Door curtain	4	
10	Pole, tent, upright, solid, 5′ 8″ long (eave pole)	10	8340-188-8405
11	End wall		
12	Line, tent, manila, both ends sewed, 5/16″ dia., 50′ long (guy line)	2	8340-252-2293
13	Line, tent, manila, one end sewed, one end w/eye, 5/16″ dia., 13′ long (door eave line)	4	8340-252-2271
14	Care and maintenance instructions flap	1	
15	Erection instructions flap	1	
16	Wall corner slide fastener	4	
17	Line, tent, manila, one end sewed, one end w/eye, 5/16″ dia., 13′ long (corner eave line)	8	8340-252-2271
18	Pin, tent, wood, 24″, or pin, steel, 12″ (for cold climate use)	28 / 28	8340-261-9751 / 8340-261-9748
19	Line, tent, cotton, footstop, ¼″ dia., 19″ long (footstop)	52	8340-252-2299
20	Pin, tent, wood, 16″, or pin, aluminum, 9″ (for cold climate use)	48 / 48	8340-261-9750 / 8340-261-9749
21	Window blackout flap	6	
22	Line, tent, manila, one end sewed, one end w/eye, 5/16″ dia., 13′ long (side eave line)	12	8340-252-2271
23	Windowpane	6	
24	Line, wall, 13½″ (tie tape)	60	
25	Side wall		
26	Window screen	6	
27	Slip, tent line, steel	28	8340-223-8094
	Line, tent, manila, one end sewed, ¼″ dia., 3′ 4″ long (eave jumper line)	14	8340-252-2291
	Line, tent, manila, one end sewed, ¼″ dia., 6′ long (ridge jumper line)	2	8340-252-2286
	Cover, tent, FWWMR, general purpose, medium	1	8340-262-2397
	Line, tent, manila, one end sewed, one end w/eye, 5/16″ dia., 13′ long (cover tie line)	2	8340-252-2271

Figure 25. Tent, general purpose, medium, FWWMR, OD, complete with pins and poles, Federal Stock No. 8340-262-2401

the inside to each side of each door entrance. When in use, the door screens are pulled across the door entrances and secured in place by tying tie tapes at the top of the screens to metal rings at the eave above the door entrances. When not in use, the door screens are rolled to the side inside the tent and secured by tying the tie tapes along the sides of the screens.

f. Windows. There are 3 window assemblies on each side of the tent below the eave. Each window assembly consists of a plastic window screen, a vinyl plastic windowpane, and a canvas blackout flap. The window screen is attached to the side wall. The windowpane is attached at the top to the side wall and secured at the bottom and two sides by a slide fastener which may be unfastened to allow the windowpane to be rolled up and tied at the top with tie tapes. A blackout flap is attached at the top to the side wall. When the flap is in use, it is secured by tying tie tapes at the two sides; when not in use, it is rolled up and tied at the top with tie tapes.

g. Ventilation.

(1) The tent is ventilated by 2 ventilators, one at the top of each end section near the ridge. The openings are protected by canvas flaps.

(2) When stoves are not being used, the stovepipe openings may also be used as ventilators.

(3) Additional ventilation can be obtained by rolling up the window blackout flaps and the windowpanes and tying them open with tie tapes.

(4) The door curtains can be opened for more ventilation.

(5) Still more ventilation can be obtained by rolling up the sides of the tent to the eaves and tying them with tie tapes.

h. Heating The tent is heated by two M-1941 tent stoves. There are 2 stovepipe openings built in the top near the 2 large upright poles of the tent. The openings are protected by canvas flaps.

i. Cover. The tent is provided with a cover for use when in storage or when being transported.

j. Liner. A liner with cover (Federal Stock No. 8340-162-2402) is available as a separate item of issue. It provides insulation from the cold in winter and reduces radiation from the sun in summer. The liner weighs 90 pounds and has a cubage in storage of 8 cubic feet. The liner may be attached to the inside of the tent at the eaves by webbing straps with buckles; to the ridge pole and hardware at the ridge, by suspension ropes with snaps; and to the center poles, by tie tapes. The liner has both fabric and screening side walls below the eaves. The fabric side walls are made of 5.2-ounce cotton cloth. The screening side walls are made of plastic. The fabric side walls may be rolled up to the eaves and secured by tie tapes and thus permit the use of the screening alone. The screening provides protection from insects and permits the liner to be used in hot as well as cold weather. There are 2 built-in ventilator screens corresponding in location to the 2 ventilators in the tent. There are 3 vinyl plastic windows on each fabric side wall corresponding in location to the windows in the tent. There are 2 stovepipe openings in the liner corresponding in location to the stovepipe openings in the tent. The liner has the following component parts:

Part	Quantity	Federal Stock No.
Line, tent, cotton, footstop, dia., 19″ long (footstop).	34	8340-252-2299
Cover, liner, tent, FWWMR, general purpose, medium.	1	No stock number available
Line, tent, manila, one end sewed, one end w/eye, 5/16″ dia., 13′ long (cover tie line).	2	8340-252-2271

29. Ground Plan

Before pitching the tent, study the ground plan carefully (fig. 26).

30. Pitching

a. Pitching Tent. Four men can pitch the tent in approximately 40 minutes.

(1) *Spreading tent out, closing slide fasteners and doors, driving 16-inch corner pins, and attaching corner footstops to pins (1, fig. 27).*

(a) Spread tent out on ground into position with corners square.

(b) Close slide fasteners at corners.

(c) Close doors by sliding door curtains across door entrances and fastening loops to wooden toggles.

(d) Drive a 16-inch wood pin or, in cold climate, 9-inch aluminum pin, at each corner, and attach corner footstops to pins.

(2) *Driving 24-inch pins, attaching eave lines, and placing eave and door poles in position for erection (2, fig. 27).*

(a) Drive the 24-inch wood pins or, in cold climate, the 12-inch steel pins, according to ground plan, using 5-foot 8-inch poles to measure distance out from tent.

(b) Attach side, corner, and door eave lines loosely to pins.

(c) Place the 5-foot 8-inch side and corner poles and the 6-foot 2-inch door eave poles in position for erection.

(3) *Inserting eave poles through grommets, raising tent walls, and running ridge pole and center poles to inside center (3, fig. 27).*

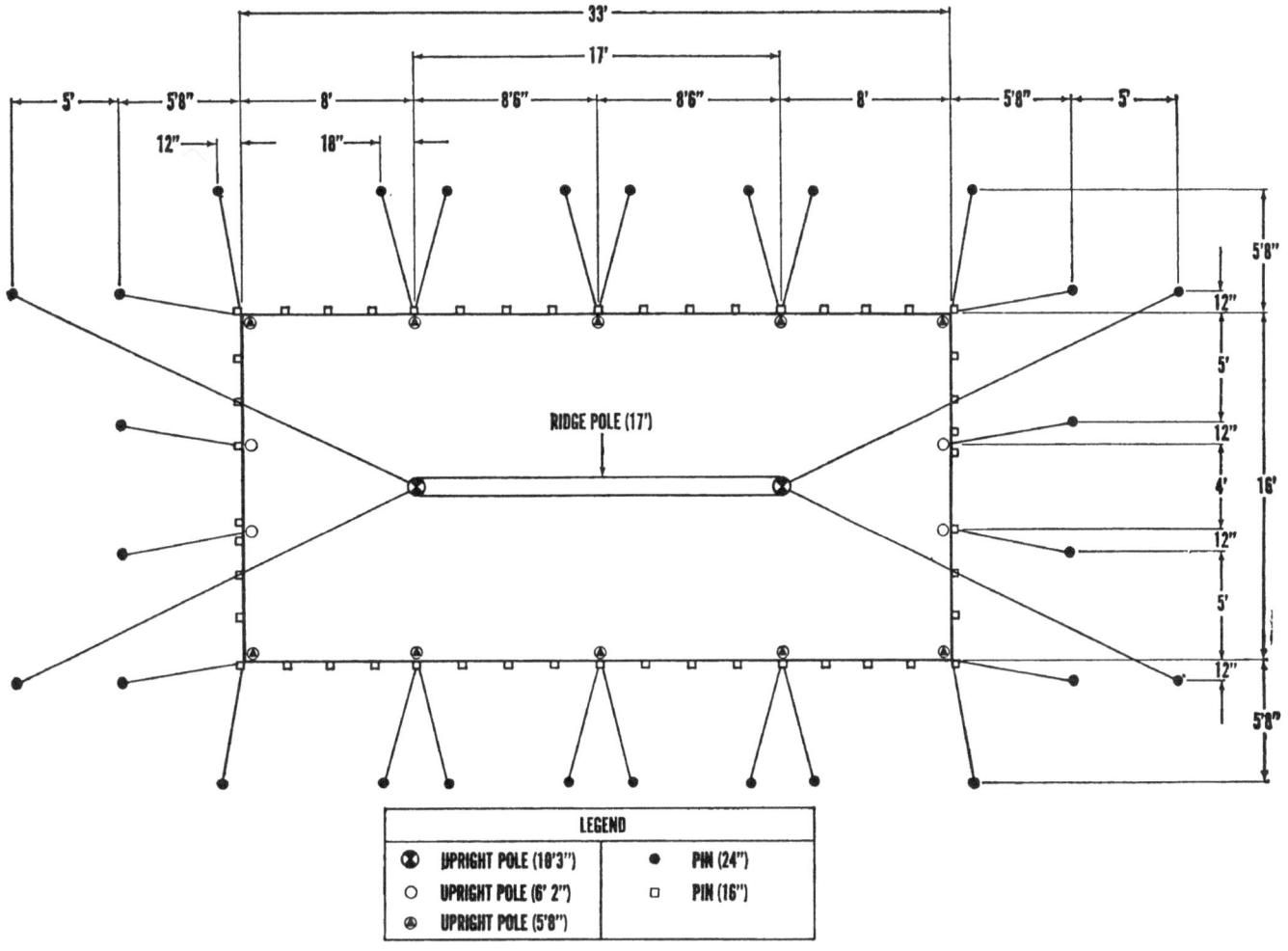

Figure 26. Ground plan of tent, general purpose, medium, FWWMR, OD.

(a) Insert spindles of the 5-foot 8-inch eave poles through eave grommets at sides and corners of tent.

(b) Insert spindles of the 6-foot 2-inch door eave poles through eave grommets at front and rear doors.

(c) Raise tent walls by raising side, corner, and door eave poles to an upright position.

(d) Tighten eave lines just enough to hold poles up.

(e) Assemble the 17-foot ridge pole and the 10-foot 3-inch center poles and slide them inside to center of tent.

(4) *Connecting center poles to ridge pole, placing spindles through rings in tent ridge, and attaching guy lines (4, fig. 27).*

(a) Connect the 10-foot 3-inch center poles to the 17-foot ridge pole by inserting spindles of center poles through holes in ridge pole, with round side of ridge pole up.

(b) Place spindles of center poles through holes in ridge plates and through rings in tent ridge.

(c) Attach guy lines to spindles of center poles outside tent before raising tent roof.

(5) *Raising center poles and attaching guy lines to pins (5, fig. 27).*

(a) With two men at each center pole, raise center poles to a vertical position.

(b) Attach guy lines to pins and tighten.

(6) *Tying jumper lines, adjusting ventilators, driving 16-inch pins, and attaching footstops to pins (6, fig. 27).*

(a) Tie jumper lines to side and corner eave poles, door eave poles, and center poles.

(b) Adjust ventilator flap lines and tie them to spindles of corner eave poles.

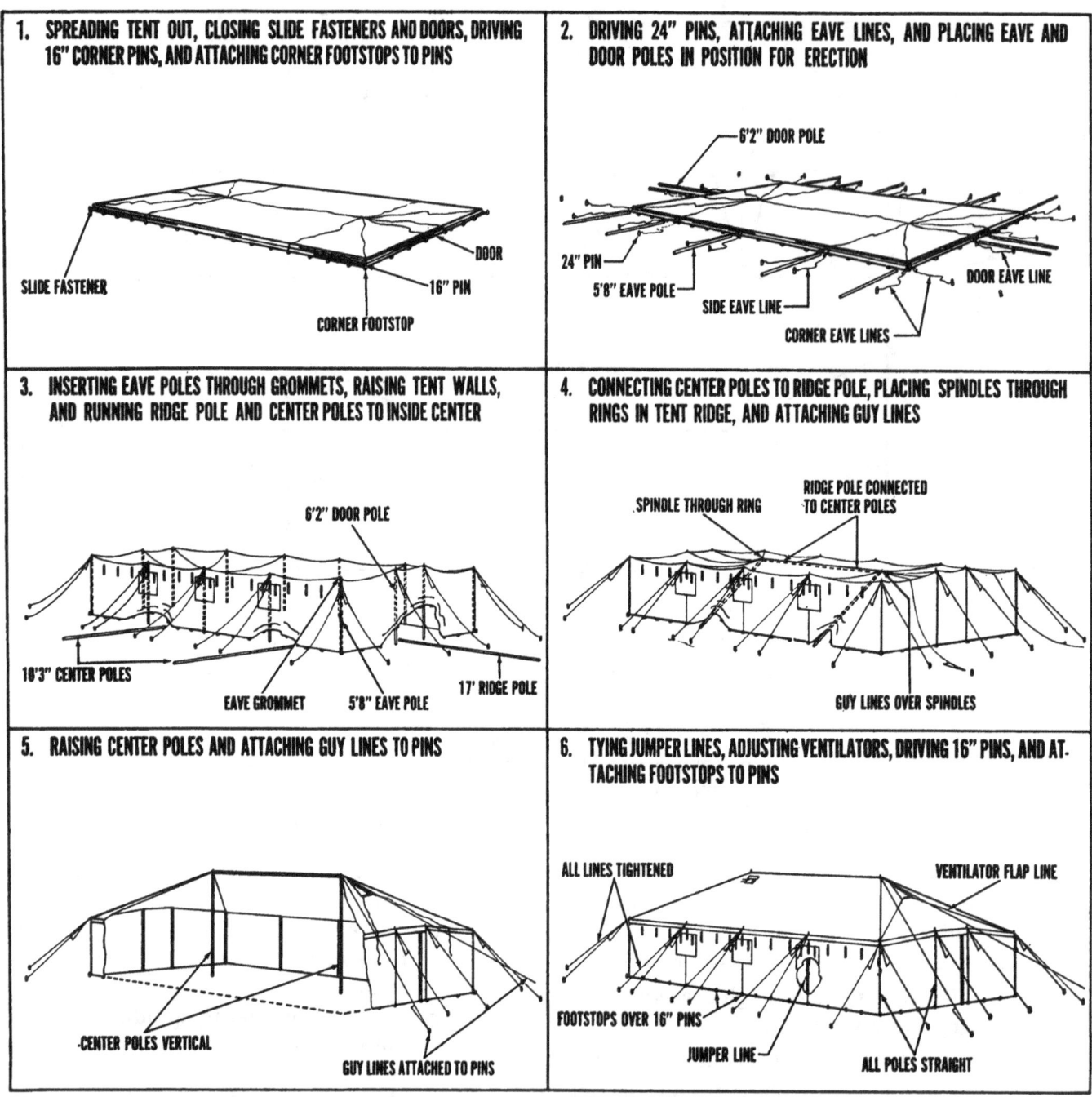

Figure 27. Steps in pitching tent, general purpose, medium, FWWMR, OD.

(c) Drive the remaining 16-inch wood pins or, in cold climate, the 9-inch aluminum pins, and attach footstops to pins.

(d) Straighten all poles.

(e) Tighten all lines and adjust tent slips until tent is smooth.

(f) Tie tie tapes at inside corners of tent around corner eave poles.

b. *Attaching Liner to Tent.*

(1) *Loosening tent lines* (1, fig. 28). Loosen slightly all tent lines by adjusting tent slips.

(2) *Unrolling liner inside tent* (2, fig. 28). Unroll liner inside tent on one side of center poles, with stovepipe openings of liner matching those of tent.

(3) *Placing liner around center poles and spreading liner* (3, fig. 28). Lift each center pole and pull liner under base of pole until hole in liner is directly under

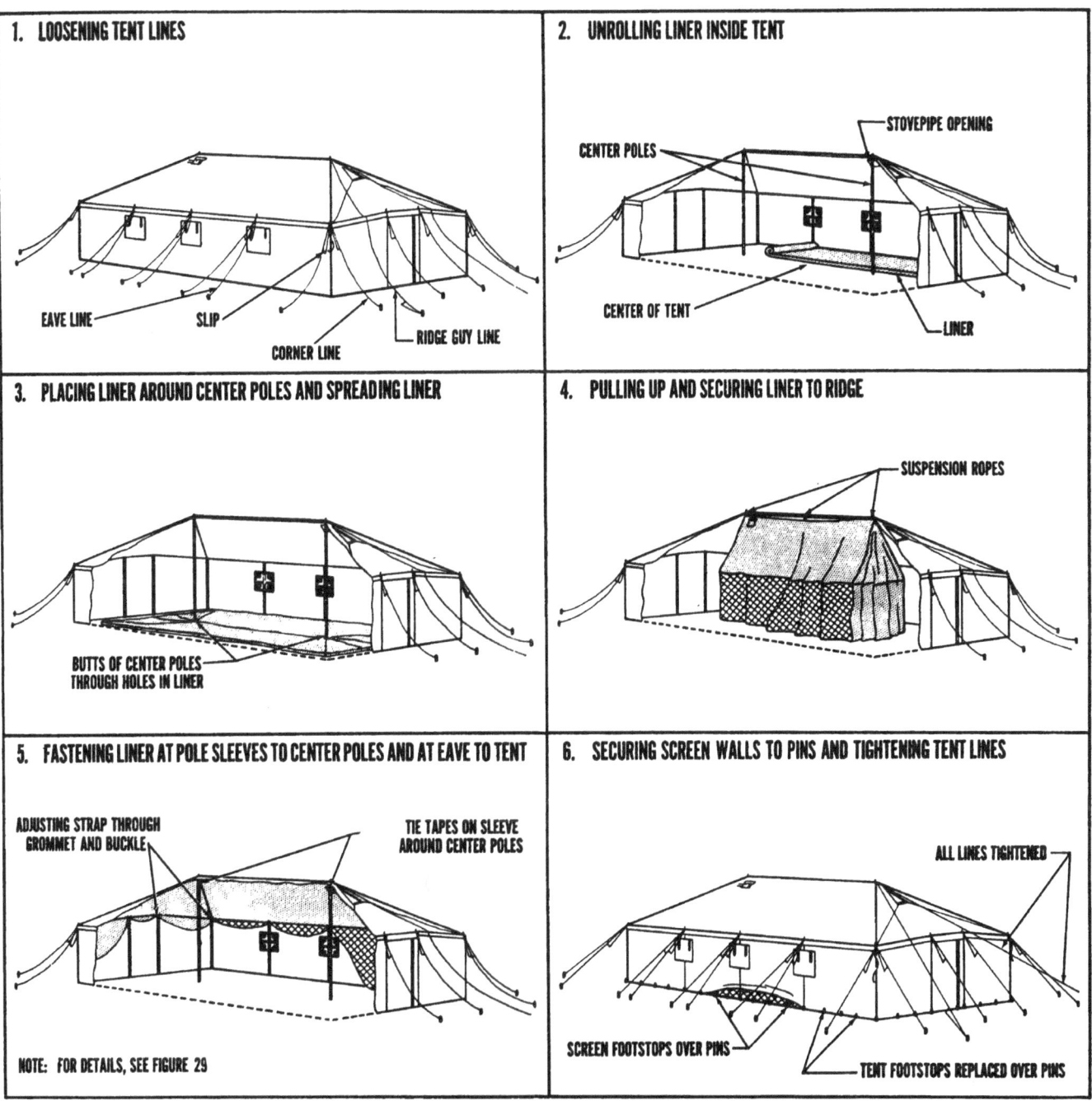

Figure 28. Steps in attaching liner to tent, general purpose, medium, FWWMR, OD.

pole; then set pole down and spread liner into position.

(4) *Pulling up and securing liner at ridge* (4, fig. 28).

  (a) Raise liner at center and run middle suspension line over ridge pole at center; insert end of suspension line through eye in other end of suspension line, and tie.

  (b) Raise liner at poles and tie ridge suspension lines to hardware at top of poles.

(5) *Fastening liner at pole sleeves to center poles and at eave to tent* (5, fig. 28).

  (a) Fasten liner at pole sleeves by tying tie tapes around center poles near the top.

  (b) Fasten liner at eave (fig. 29) by running each suspension line (attached to outside of liner) around hardware (square with hook or diamond with hook) or through webbing on inside of tent. Then run suspension line through the No. 4 oblong grommet to

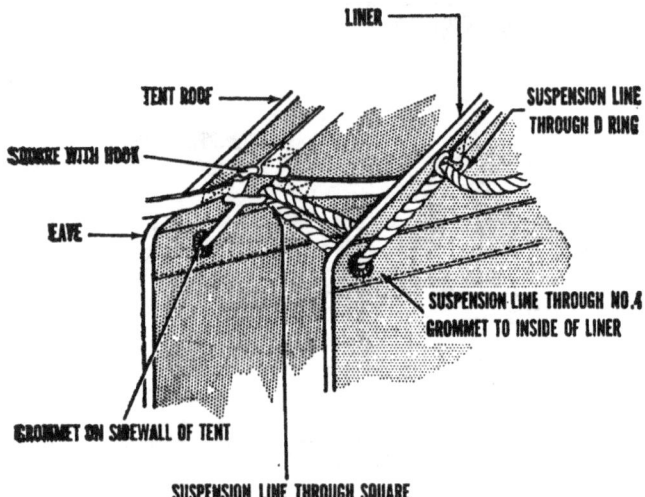

Figure 29. Fastening liner to eave of tent, general purpose, medium, FWWMR, OD.

inside of liner, tighten suspension line so that there is a 4-inch space between tent and liner, and secure suspension line to **D**-ring inside.

(c) Tie together tie tapes at corners of liner. Tie tie tapes at door entrances around door eave poles.

(6) *Securing screen walls to pins and tightening tent lines* (6, fig. 28).

(a) Secure screen walls of liner by removing tent footstops from the 16-inch wood or the 9-inch aluminum pins, placing screen footstops on pins, and replacing tent footstops on pins.

(b) Tighten all tent lines by adjusting tent slips.

31. Striking

Four men can strike the tent in approximately 30 minutes.

a. *Removing Liner.*
(1) Unfasten footstops of tent and liner screen from the 16-inch wood or the 9-inch aluminum pins.
(2) Untie tie tapes at corners. Untie tie tapes at door entrances from door eave poles.
(3) Unfasten suspension lines from **D**-rings at eave, pull lines through the No. 4 oblong grommets to outside of liner, and unfasten line from webbing of tent or from hardware.
(4) Untie tie tapes from around center poles at top.

(5) Unfasten suspension lines from hardware at ridge and from around ridge pole.
(6) Lift center poles slightly and remove liner from tent.

b. *Striking Tent.*
(1) Close the 4 corner slide fasteners. Close doors and fasten wooden toggles to toggle loops.
(2) Remove all but the 4 corner 16-inch footstop pins.
(3) Unfasten from the 24-inch wood or the 12-inch steel pins door eave lines and all other eave lines except those at corners. Remove the 24-inch wood or 12-inch steel pins from which door and side eave lines were unfastened.
(4) Remove door eave poles and all eave poles except those at corners.
(5) With a man at each center pole and a man outside the tent at each end, remove guy lines from the 24-inch wood pins, or the 12-inch steel pins, and lower center poles gently to ground. Remove the 24-inch wood or the 12-inch steel guy-line pins.
(6) Disconnect center poles from ridge pole and remove poles from tent. Disassemble ridge pole and center poles.
(7) Unfasten from the 24-inch wood or the 12-inch steel pins the 8 corner eave lines. Unfasten tie tapes from around corner eave poles and remove poles.

32. Folding

a. *Folding Liner.*
(1) *Spreading liner out flat, folded at ridge, and with end sections folded inside* (1, fig. 30). Spread liner flat on ground and fold at ridge, one half folded over the other. Tuck ends neatly inward toward center with an accordion fold. Coil suspension ropes down toward center and adjusting straps up toward ridge.
(2) *Folding ends toward center* (2, fig. 30). Fold ends toward center over body of liner.
(3) *Folding screen up at eave* (3, fig. 30). Fold screen up at eave. Coil adjusting straps toward center.

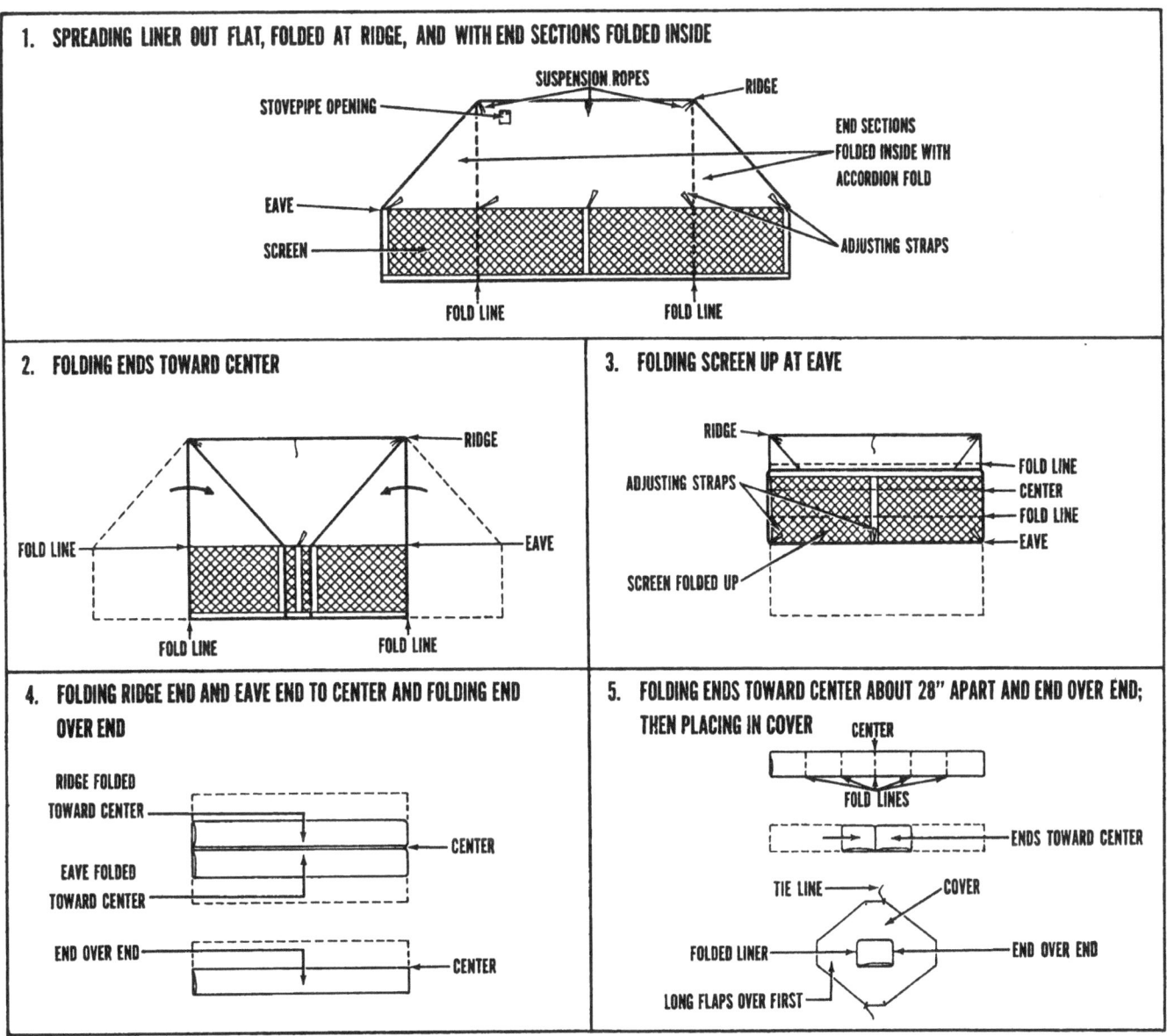

Figure 30. Folding tent liner, FWWMR, general purpose, medium.

(4) *Folding ridge end and eave end to center and folding end over end* (4, fig. 30). Fold ridge end and eave end to center and fold end over end.

(5) *Folding ends toward center about 28 inches apart and end over end; then placing in cover* (5, fig. 30). Fold ends toward center in folds of about 28 inches each and fold end over end. Place in cover and close cover with long flaps over first, making sure flaps are neatly folded within package. Tie cover with the 2 tie lines.

b. *Folding Tent.*
(1) *Preliminary procedures* (1, fig. 31).

(a) Close and secure doors, window assemblies, and stovepipe openings.
(b) Spread tent out flat.
(c) Fold tent at ridge, one half folded over the other.
(d) Coil eave lines toward center.

(2) *Folding end walls over toward center* (2, fig. 31). Fold end walls over toward center.

(3) *Folding roof over so that ridge is at eave* (3, fig. 31). Fold roof over so that ridge is at eave.

(4) *Folding roof over again to eave and folding side walls to eave in two folds* (4, fig. 31).

(a) Fold roof over again to eave.
(b) Fold side walls to eave in 2 folds, making sure the second fold line is directly below windows.

(5) *Folding bulk of tent over side wall (5, fig. 31).* Fold bulk of tent (folded roof) over side walls, exposing eave. Then coil remaining eave lines on top of folded tent.

(6) *Folding each end toward center and inserting guy lines (6, fig. 31).* Fold each end toward center in 3 folds, leaving 1 foot in the middle. Then, insert in the middle the coiled and tied guy lines.

(7) *Folding end over end and placing in cover (7, fig. 31).* Fold end over end and place in cover. Close cover with long flaps over first, making sure flaps are neatly folded within package. Tie cover with the two tie lines.

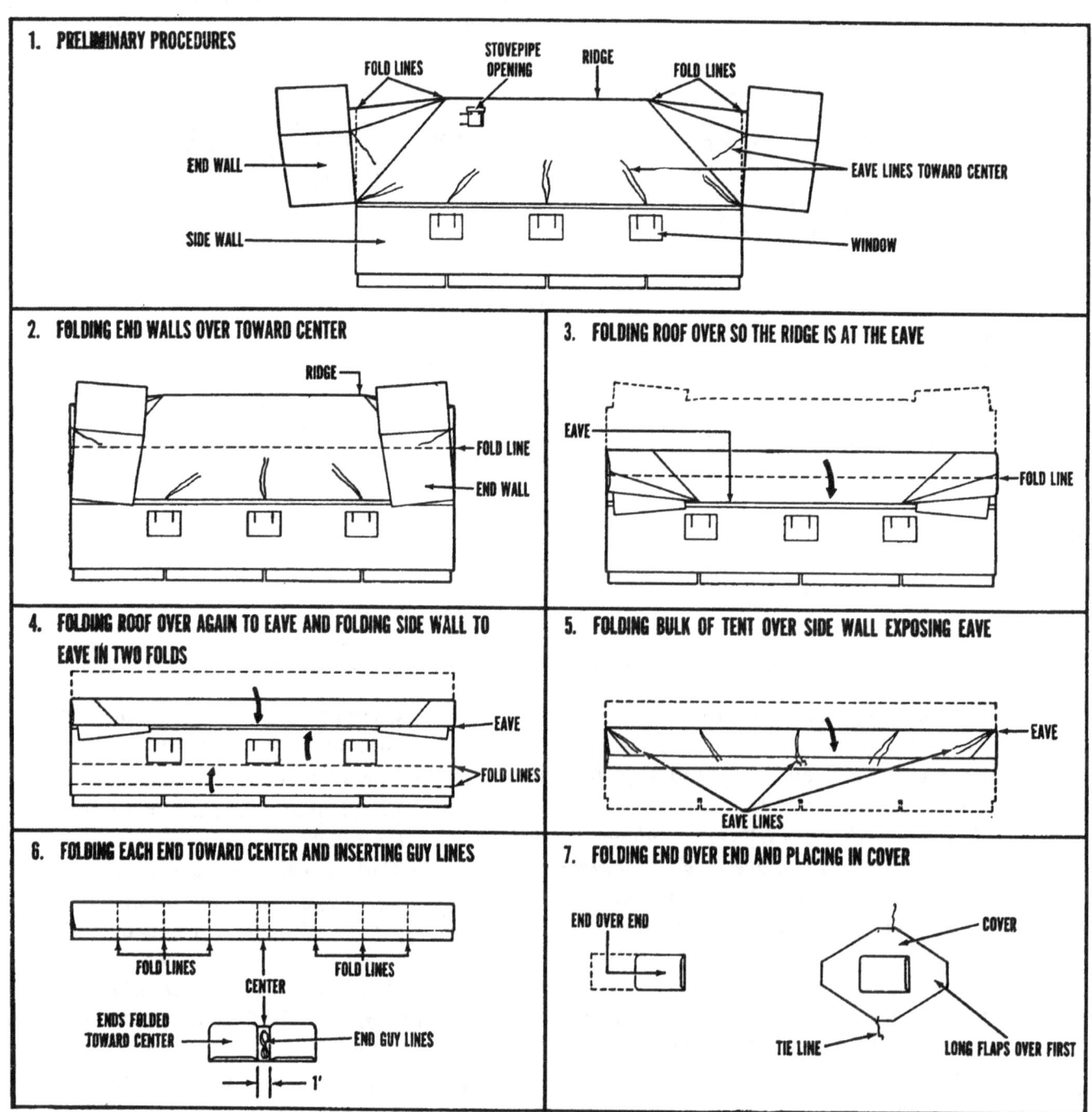

Figure 31. Folding tent, general purpose, medium, FWWMR, OD.

Section VI. HEXAGONAL TENT

33. Use

The tent, hexagonal, lightweight, M-1950, FWWMR, OD, complete with pins and poles (fig. 32), is used to provide shelter for troops operating in extremely cold or cold-wet areas. The tent will normally accommodate 5 men and their individual clothing and equipment; under emergency conditions, the tent will provide shelter for 5 men sleeping and 1 on watch.

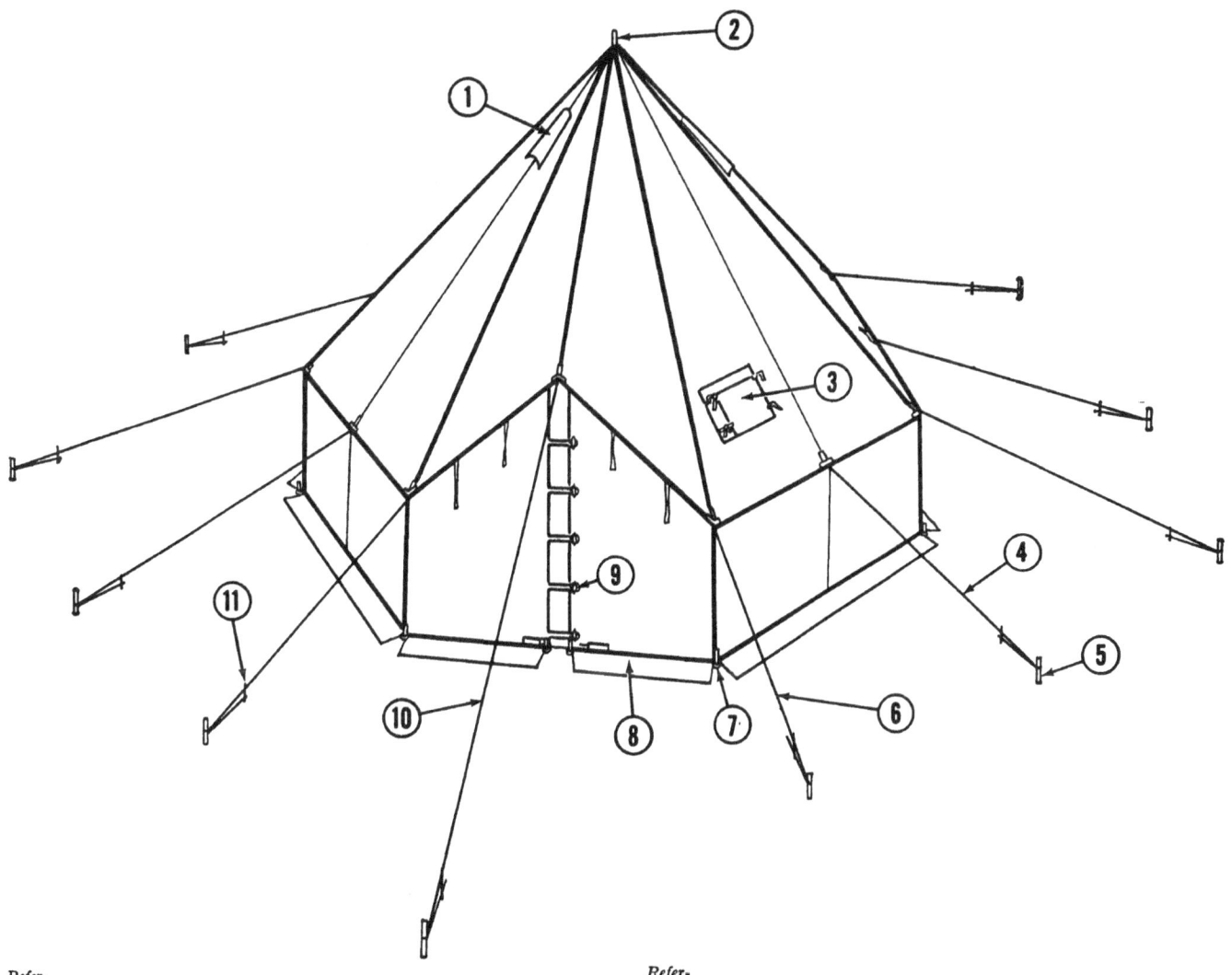

Reference No.	Part	Quantity	Federal Stock No.
1	Ventilator	2	
2	Pole, tent, telescopic, upright, jointed, magnesium, 4' 9" to 9'	1	8340-188-8413
3	Stovepipe opening	1	
4	Line, tent, cotton, 12' 6", unfinished, 7/32" dia. (intermediate eave line)	5	8304-252-6912
5	Pin, tent, aluminum, 9"	20	8340-261-9749
6	Line, tent, cotton, 12' 6", unfinished, yellow, 7/32" dia. (corner eave line)	6	8340-262-3658
7	Line, tent, cotton, footstop, 1/4" dia. 19" long (footstop)	8	8340-252-2299
8	Snow cloth		
9	Toggle, tent, wood	5	8340-376-1295
10	Line, tent, cotton, 21' 6", unfinished, 7/32" dia. (door eave line)	1	8340-252-6913
11	Slip, tent line, magnesium	12	8340-223-8095
	Line, tent, cotton, 19', unfinished, 1/8" dia. (sock line)	1	8340-252-6911
	Line, tent, cotton, 30', unfinished, 1/8" dia. (sock line)	1	8340-252-2304
	Line, tent, cotton, 35', unfinished, 1/4" dia. (sock line)	1	8340-252-6910
	Toggle, tent, steel	36	8340-242-7872
	Tent liner, FMR, hexagonal, lightweight, M-1950	1	8340-262-3700
	Cover, tent, hexagonal, lightweight, M-1950	1	8340-241-8435

Figure 32. Tent, hexagonal, lightweight, M-1950, FWWMR, OD, complete with pins and poles, Federal Stock No. 8340-269-1372.

34. Description

The tent is a six-sided, pyramidal tent, supported by a telescopic pole.

a. Dimensions. Each side of the tent is 6 feet 7½ inches long. The tent is 8 feet 6 inches high at the peak. The wall height is 2 feet, giving a pitch of 6 feet 6 inches. The hexagonal floor of the tent is approximately 13 feet 3 inches in diameter.

b. Weight and Cubage. The tent weighs 40 pounds, and the pins and pole weigh 8 pounds. The tent has a cubage in storage of 3.6 cubic feet, and the pins and pole have a cubage of 0.2 cubic feet.

c. Floorspace. The floorspace of the tent is 113.2 square feet.

d. Material. The tent is made of plied yarn, wind-resistant, sateen cotton cloth with fire, water, and mildew resistant treatment, and weighs approximately 9 ounces per square yard.

e. Door. The tent has 1 door 5 feet high which is located in the center of one side. Door flaps may be closed either by the slide fastener or by loops over wood toggles.

f. Ventilation. The tent is ventilated by 2 built-in ventilators on opposite sides and near the peak of the tent. The ventilators have inside ducts, which may be closed by tie cords. The ventilator hoods are of the fixed type, each hood being made with a stiffener inserted in the hem to keep it extended out from the ventilator opening.

g. Heating. The tent is heated by an M-1950 Yukon stove. A stovepipe opening with a silicone rubber-molded ring is built in one side of the tent near the eave. When the stove is not in use, the stovepipe opening may be protected by a canvas flap.

h. Sock Lines. Three sock lines are provided for drying clothing and equipment.

i. Snow Cloths. There is a snow cloth sewed to the bottom of each side of the tent. When the tent is pitched, the snow cloths are flat on the ground on the outside of the tent. Snow is deposited on the snow cloths for insulation purposes.

j. Liner. A fire-resistant liner, made of 5.2-ounce cotton cloth, is provided to insulate the tent and to prevent frost from falling on the occupants. The liner is held in place by metal toggles.

k. Cover. The tent is provided with a cover for use when in storage or when being transported. The tent and liner, when folded, fit into the cover. Aluminum tent pins are nested, and the magnesium pole is telescoped to its shortest length and placed in the pocket at one side of the cover.

35. Ground Plan

Before pitching the tent, study the ground plan carefully (fig. 33).

36. Pitching

The tent can be pitched by 5 men in approximately 15 minutes.

a. Preliminary Procedures (1, fig. 34).
  (1) Spread tent on ground. Check to see if liner is in place; usually, it is not in place in a new tent. If liner is not in place, spread it out beneath tent.
  (2) Secure **D**-ring to snap inside door.
  (3) Close slide fastener in door.
  (4) Drive 6 corner pins and 2 door pins and attach footstops to pins.

b. Attaching Corner Eave Lines and Inserting Tent Pole (2, fig. 34).
  (1) Drive 6 pins about 6 feet from corners of tent and attach yellow corner eave lines to pins. Pins on opposite sides of tent should be in a straight line.
  (2) Open door and push pole, extended to 8'6", under tent.
  (3) Insert spindle of pole through grommet in peak of liner and through hand-worked ring in peak of tent.

c. Raising Tent (3, fig. 34).
  (1) With one man inside the tent, close inside and outside **D**-rings and snaps on door; close slide fastener.
  (2) Fasten loops to wood toggles on door.
  (3) Raise tent and liner; place butt of tentpole in center of tent area.

d. Attaching Door Eave Line and Intermediate Eave Lines (4, fig. 34).
  (1) Stake door eave line far enough to hold door vertical.
  (2) Stake intermediate eave line pins.
  (3) Attach the 5 intermediate eave lines to pins.
  (4) Adjust and tighten all lines.

e. Propping Up Door Eave Line. The door eave line may be propped up by placing the line over an improvised pole (tree branch or other object higher than the door entrance) at a distance of about 5 feet in front of the door and then staking the line out to a pin. This keeps the door from

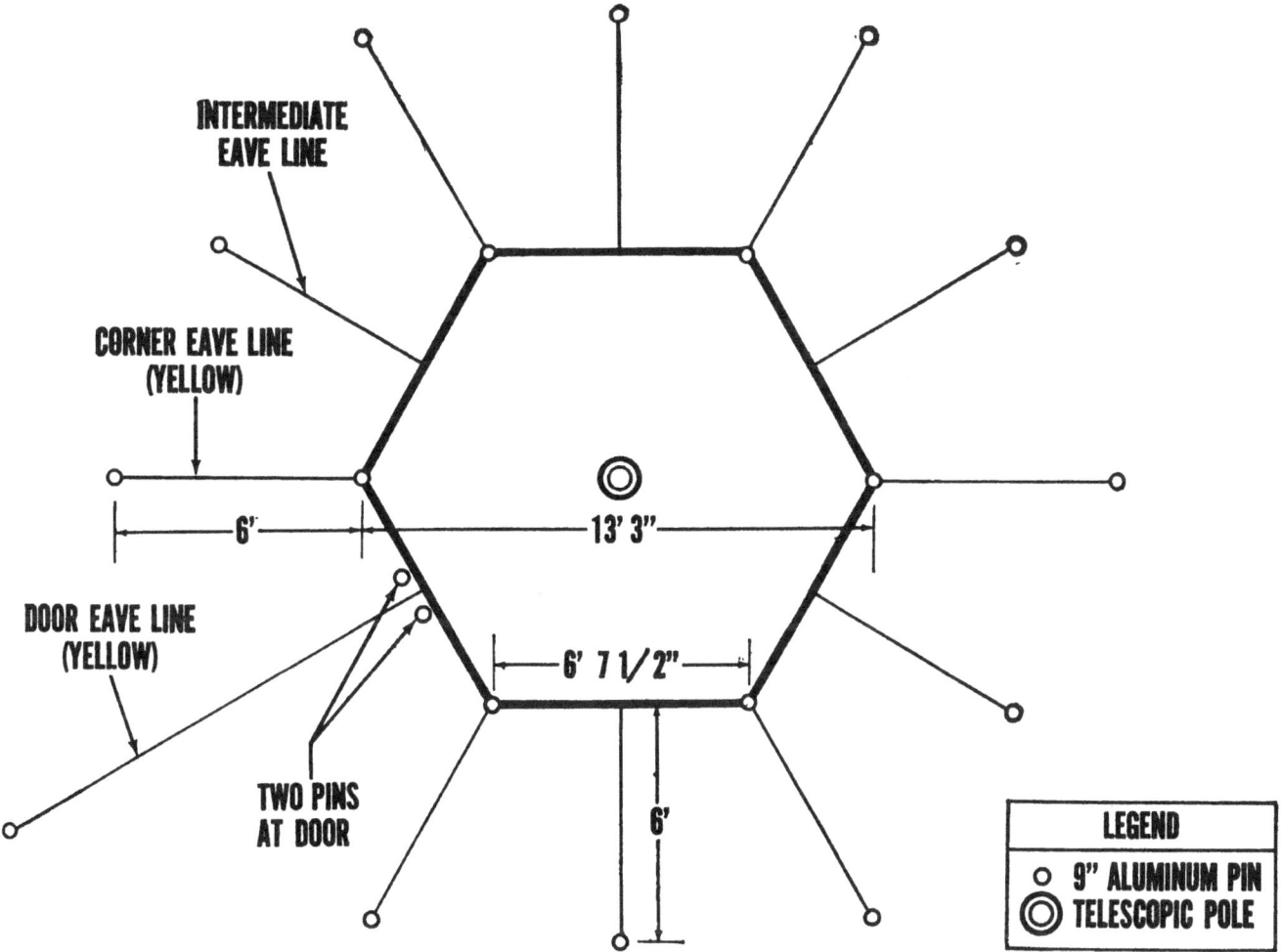

Figure 33. Ground plan of tent, hexagonal, lightweight, M-1950, FWWMR, OD.

sagging, makes the slide fastener work better, makes the tent easier to get into and out of, and gives the tent greater stability.

f. Fastening Liner. Fasten liner in place by inserting wire toggles, which are secured to tent, into grommets set in liner. Allow approximately 2 inches between tent and liner for insulating purposes. Tie tapes around stovepipe opening in liner to corresponding tapes around stovepipe opening in tent to keep stovepipe opening in place. The 35-foot sock line is threaded through the eyes of the toggles at the eave line and tied to the toggle at each corner of the door. The 30-foot sock line is threaded through the eyes of the next row of toggles and the two ends are tied in a square knot. The 19-foot sock line is threaded through the eyes of the remaining row of toggles and the ends are tied in a square knot.

37. Striking

a. Loosen liner tie tapes.
b. Loosen door eave line.
c. Remove all footstop pins.
d. Loosen all eave lines and remove all eave-line pins.
e. Remove tent pole, and telescope pole to its shortest length.
f. Remove liner only if repairs are needed.

38. Folding

a. Folding Tent.
   (1) *Engaging snap into D-ring and closing slide fastener.* Engage snap into D-ring inside door and close door slide fasteners.
   (2) *Spreading tent out to fold* (1, fig. 35). Spread tent on ground and locate stovepipe opening panel on top fold. Grasp corner eave line (to the right of stovepipe opening) and pull out corner of panel. Then coil intermediate eave line neatly on extended panel.
   (3) *Making first panel fold* (2, fig. 35). Reaching to the left, grasp corner eave line (to the left of stovepipe opening)

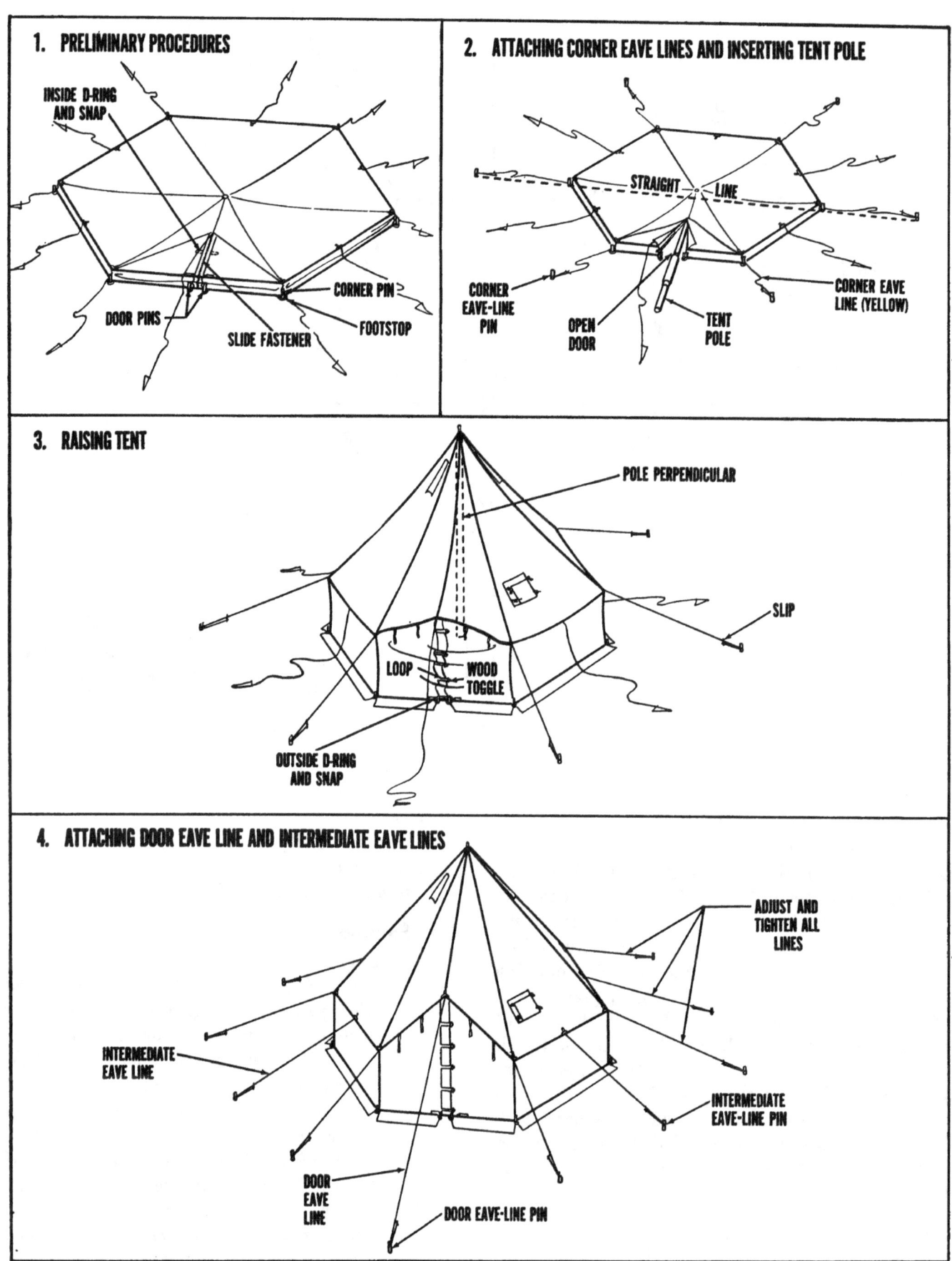

Figure 34. Steps in pitching tent, hexagonal, lightweight, M-1950, FWWMR, OD.

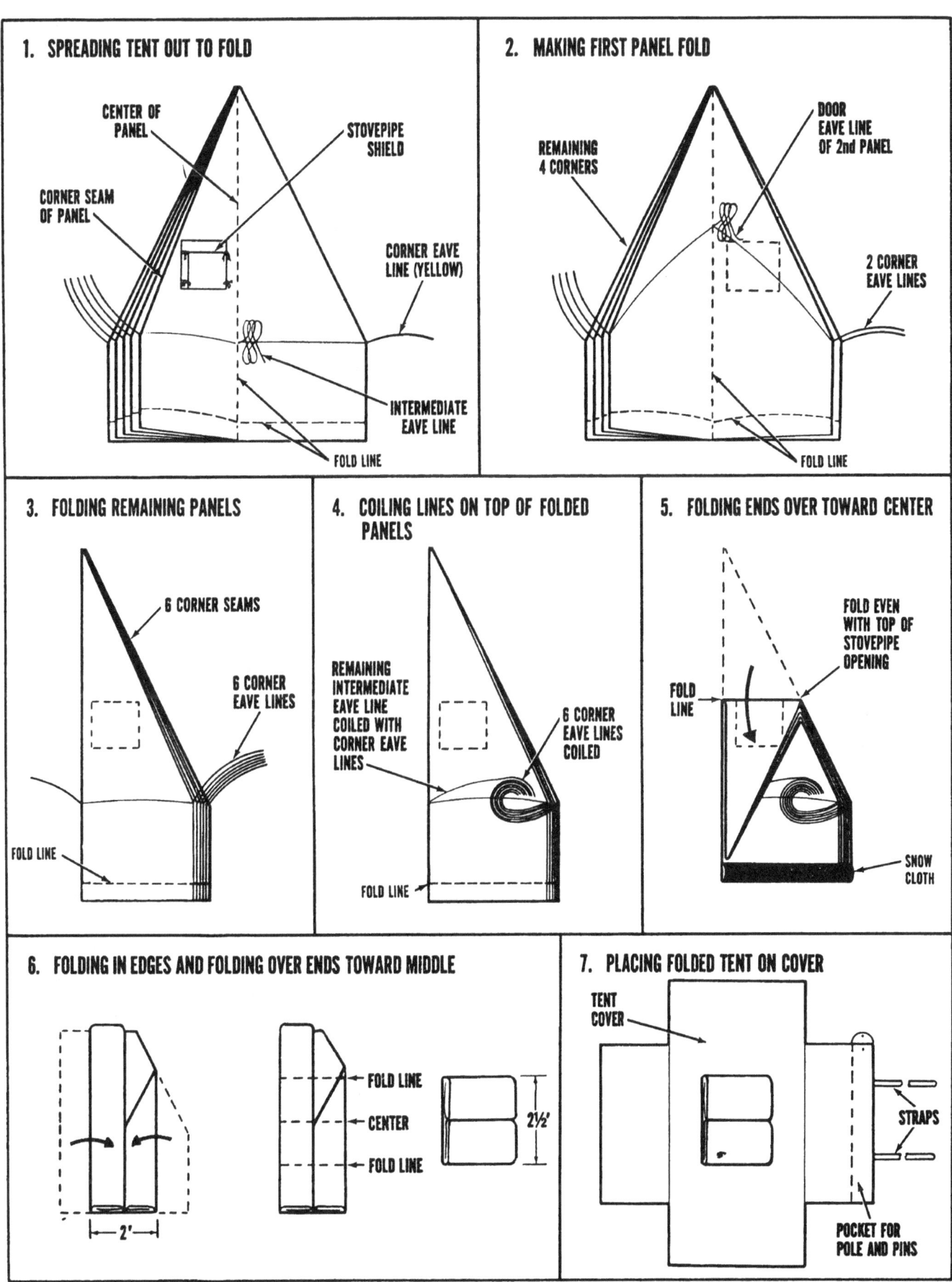

Figure 35. Steps in folding tent, hexagonal, lightweight, M-1950, FWWMR, OD.

and pull the second panel to the right, making an accordion fold.

(4) *Folding remaining panels* (3, fig. 35). Fold remaining panels in the same manner, having 6 folds in all. As each fold is completed, coil intermediate eave lines or door eave line neatly between folds.

(5) *Coiling lines on top of folded panels* (4, fig. 35). Coil on top of folded tent panels the 6 corner eave lines that have been drawn to the right and the last remaining intermediate eave line.

(6) *Folding ends over toward center* (5, fig. 35). Fold snow cloth up over side walls of tent. Fold ends over toward center, even with top of stovepipe opening.

(7) *Folding in edges and folding over ends toward middle* (6, fig. 35). Fold in edges and fold over edges toward middle.

(8) *Placing folded tent on cover* (7, fig. 35). Place folded tent on cover; place nested pins and telescoped pole into pocket of cover.

(9) *Closing cover.* Close cover, securing it with straps and loops. Care should be taken that flaps are tucked neatly inside cover.

b. *Folding Liner.* Ordinarily the liner is not removed from the tent. When necessary, the liner may be folded separately in the same manner as the tent. After folding, the liner may be placed inside the cover with the tent, pins, and pole.

Section VII. HOSPITAL TENT, SECTIONAL

39. Use

The tent, hospital, sectional, FWWMR, OD, complete with pins and poles (fig. 36), is used by field hospital units as a hospital ward or as a surgical operating room. The tent, with 2 end sections and 3 middle sections, has a capacity of 24 cots. Five additional cots may be accommodated with each middle section added.

40. Description

The tent is a hip-roofed, square-end tent, made in sections. The tent normally consists of 2 end sections and 3 middle sections, with a vestibule at each end. Since the tent is sectional in construction, it may be extended to any desired length by means of additional middle sections.

a. *Dimensions.*

(1) The tent, consisting of 2 end sections and 3 middle sections, is 18 feet wide and 53 feet long (exclusive of the vestibules). Each end section is 9 feet 3 inches long and each middle section is 12 feet long. A total of 18 inches of canvas is used as overlap in lacing the sections together.

(2) The ridge height of the tent is 12 feet and the wall height is 6 feet, giving a pitch of 6 feet.

(3) Each of the 2 vestibules is 7 feet 6 inches long and 4 feet wide. The vestibules are 8 feet high where they adjoin the tent and 6 feet high at the entrances.

b. *Weight and Cubage.* The tent weighs 770 pounds, and the pins and poles weigh 327 pounds. The tent has a cubage in storage of 31.5 cubic feet, and the pins and poles have a cubage of 12.2 cubic feet.

c. *Floorspace.* The floorspace is 954 square feet in the main part of the tent and 60 square feet in the 2 vestibules.

d. *Materials.* The top, side walls, and all reinforcements are made of 12.29-ounce duck; the sod cloth is made of 9.85-ounce duck. The tent is suspended by webbing rather than canvas. The webbing carries the weight of the canvas. Steel plates are used to eliminate friction between the webbing and eave lines.

e. *Attaching Parts.* Corners and side walls, from eave to sod cloth, are held together by snap fasteners. Tent top sections are laced together by web straps slipped underneath metal loops that have been pushed through oblong grommets. Two water flaps overlap the lacing between sections and prevent leakage in rainy weather. The water flaps are held in place by tie lines.

f. *Windows.* The tent has 16 window openings (2 in each end section and 4 in each middle section), into which flexible translucent window sashes may be snapped.

g. *Ventilation.*

(1) The tent is ventilated through stovepipe openings near the ridge of each middle section and by openings near the ridge of each end section. Screen sashes are pro-

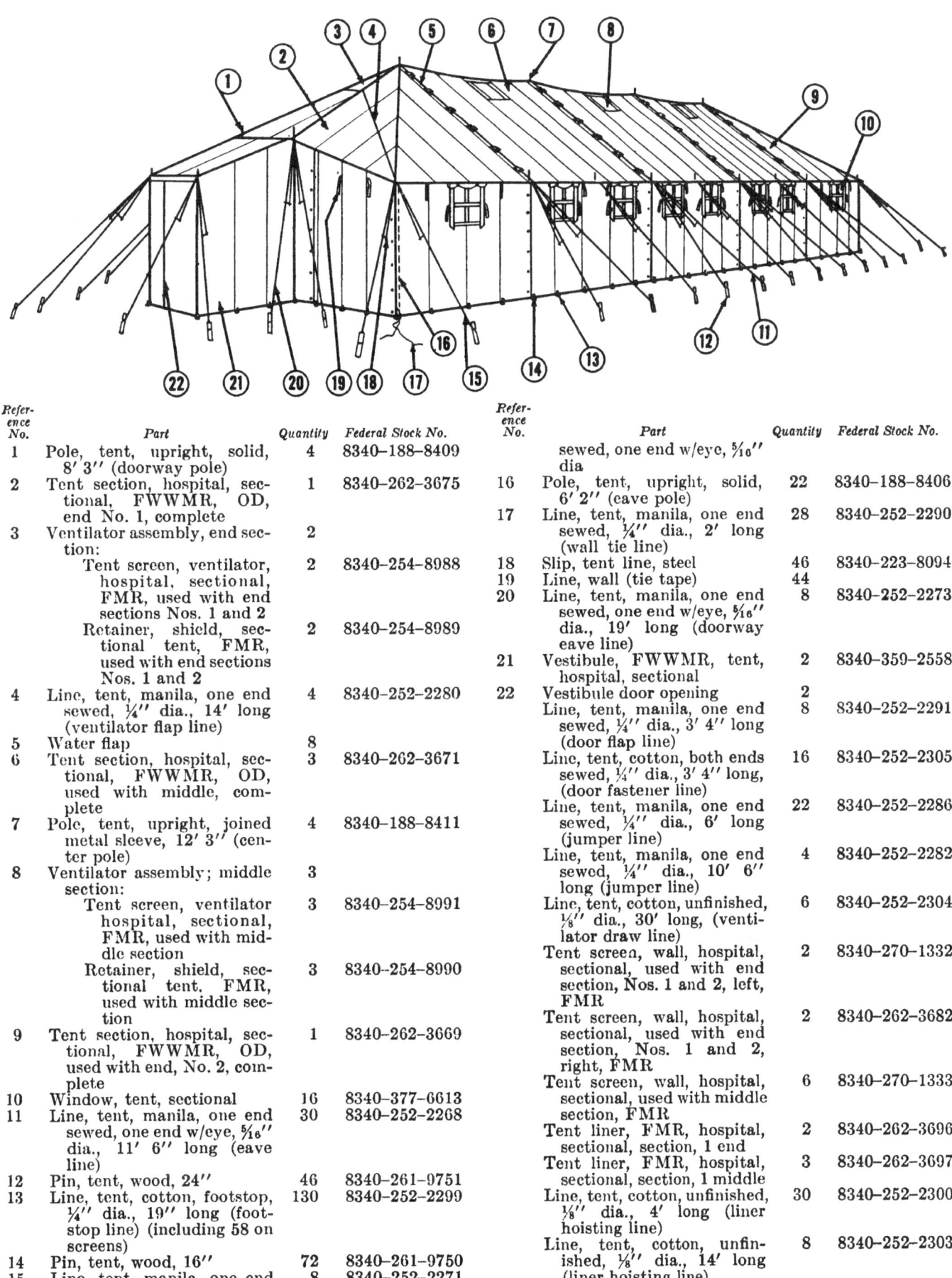

Reference No.	Part	Quantity	Federal Stock No.
1	Pole, tent, upright, solid, 8' 3" (doorway pole)	4	8340-188-8409
2	Tent section, hospital, sectional, FWWMR, OD, end No. 1, complete	1	8340-262-3675
3	Ventilator assembly, end section:	2	
	Tent screen, ventilator, hospital, sectional, FMR, used with end sections Nos. 1 and 2	2	8340-254-8988
	Retainer, shield, sectional tent, FMR, used with end sections Nos. 1 and 2	2	8340-254-8989
4	Line, tent, manila, one end sewed, ¼" dia., 14' long (ventilator flap line)	4	8340-252-2280
5	Water flap	8	
6	Tent section, hospital, sectional, FWWMR, OD, used with middle, complete	3	8340-262-3671
7	Pole, tent, upright, joined metal sleeve, 12' 3" (center pole)	4	8340-188-8411
8	Ventilator assembly; middle section:	3	
	Tent screen, ventilator hospital, sectional, FMR, used with middle section	3	8340-254-8991
	Retainer, shield, sectional tent, FMR, used with middle section	3	8340-254-8990
9	Tent section, hospital, sectional, FWWMR, OD, used with end, No. 2, complete	1	8340-262-3669
10	Window, tent, sectional	16	8340-377-6613
11	Line, tent, manila, one end sewed, one end w/eye, 5/16" dia., 11' 6" long (eave line)	30	8340-252-2268
12	Pin, tent, wood, 24"	46	8340-261-9751
13	Line, tent, cotton, footstop, ¼" dia., 19" long (footstop line) (including 58 on screens)	130	8340-252-2299
14	Pin, tent, wood, 16"	72	8340-261-9750
15	Line, tent, manila, one end sewed, one end w/eye, 5/16" dia	8	8340-252-2271
16	Pole, tent, upright, solid, 6' 2" (eave pole)	22	8340-188-8406
17	Line, tent, manila, one end sewed, ¼" dia., 2' long (wall tie line)	28	8340-252-2290
18	Slip, tent line, steel	46	8340-223-8094
19	Line, wall (tie tape)	44	
20	Line, tent, manila, one end sewed, one end w/eye, 5/16" dia., 19' long (doorway eave line)	8	8340-252-2273
21	Vestibule, FWWMR, tent, hospital, sectional	2	8340-359-2558
22	Vestibule door opening	2	
	Line, tent, manila, one end sewed, ¼" dia., 3' 4" long (door flap line)	8	8340-252-2291
	Line, tent, cotton, both ends sewed, ¼" dia., 3' 4" long, (door fastener line)	16	8340-252-2305
	Line, tent, manila, one end sewed, ¼" dia., 6' long (jumper line)	22	8340-252-2286
	Line, tent, manila, one end sewed, ¼" dia., 10' 6" long (jumper line)	4	8340-252-2282
	Line, tent, cotton, unfinished, ⅛" dia., 30' long, (ventilator draw line)	6	8340-252-2304
	Tent screen, wall, hospital, sectional, used with end section, Nos. 1 and 2, left, FMR	2	8340-270-1332
	Tent screen, wall, hospital, sectional, used with end section, Nos. 1 and 2, right, FMR	2	8340-262-3682
	Tent screen, wall, hospital, sectional, used with middle section, FMR	6	8340-270-1333
	Tent liner, FMR, hospital, sectional, section, 1 end	2	8340-262-3696
	Tent liner, FMR, hospital, sectional, section, 1 middle	3	8340-262-3697
	Line, tent, cotton, unfinished, ⅛" dia., 4' long (liner hoisting line)	30	8340-252-2300
	Line, tent, cotton, unfinished, ⅛" dia., 14' long (liner hoisting line)	8	8340-252-2303

Figure 36. Tent, hospital, sectional, FWWMR, OD, complete with pins and poles.

vided for all tent top openings, and all openings are protected by canvas flaps. The canvas flaps covering the stovepipe openings can be adjusted from the inside of the tent by draw lines which run from ventilator flaps through D-rings and bull's-eyes to within easy reach at the side walls. The end section ventilator flaps can be adjusted by exterior flap lines and held in place by hitching the lines to the upright corner pole spindles.

(2) Additional ventilation can be obtained by removing the flexible window sashes.

(3) Still more ventilation can be obtained by rolling up the canvas side walls and attaching the insect netting screens.

h. Screens. Each section is provided with insect netting screens. Screen sections are joined and held to the inside eave of the tent by snap fasteners.

i. Heating. The tent is heated by three M-1941 tent stoves when the three middle sections are used. An extra stove can be added with each additional middle section.

j. Shield Retainer Pockets. The side walls of each tent section are constructed with 10- by 20-inch tent shield retainer pockets to store the retainer shields when tent is not being heated.

k. Ballast Pockets. Vestibule and end sections are constructed with 8- by 15½-inch ballast pockets, which may be filled with sand or gravel to help hold the canvas down.

l. Liners. Each section has a liner made of 4-ounce unbleached water-repellent and mildew-resistant cotton sheeting. The liner covers the area of the tent top, hanging just below it, and has a border dropping approximately 10 inches along the side and end walls.

m. Packaging. Each tent section comes folded separately, including its necessary components, such as window sashes and insect netting screens.

41. Ground Plan

Before pitching the tent, study the ground plan carefully (fig. 37). The tent site should be at least 80 feet long and 30 feet wide in order to accomodate the tent, with its vestibules, and to allow for a 6-foot guy area.

42. Pitching

The tent can be pitched by 9 men in approximately 90 minutes.

a. Driving Corner and Side Guy-line Pins and Digging 3-inch Holes for Center Poles (1, fig. 38).

(1) Drive corner and side guy-line (24-inch) pins according to ground plan.

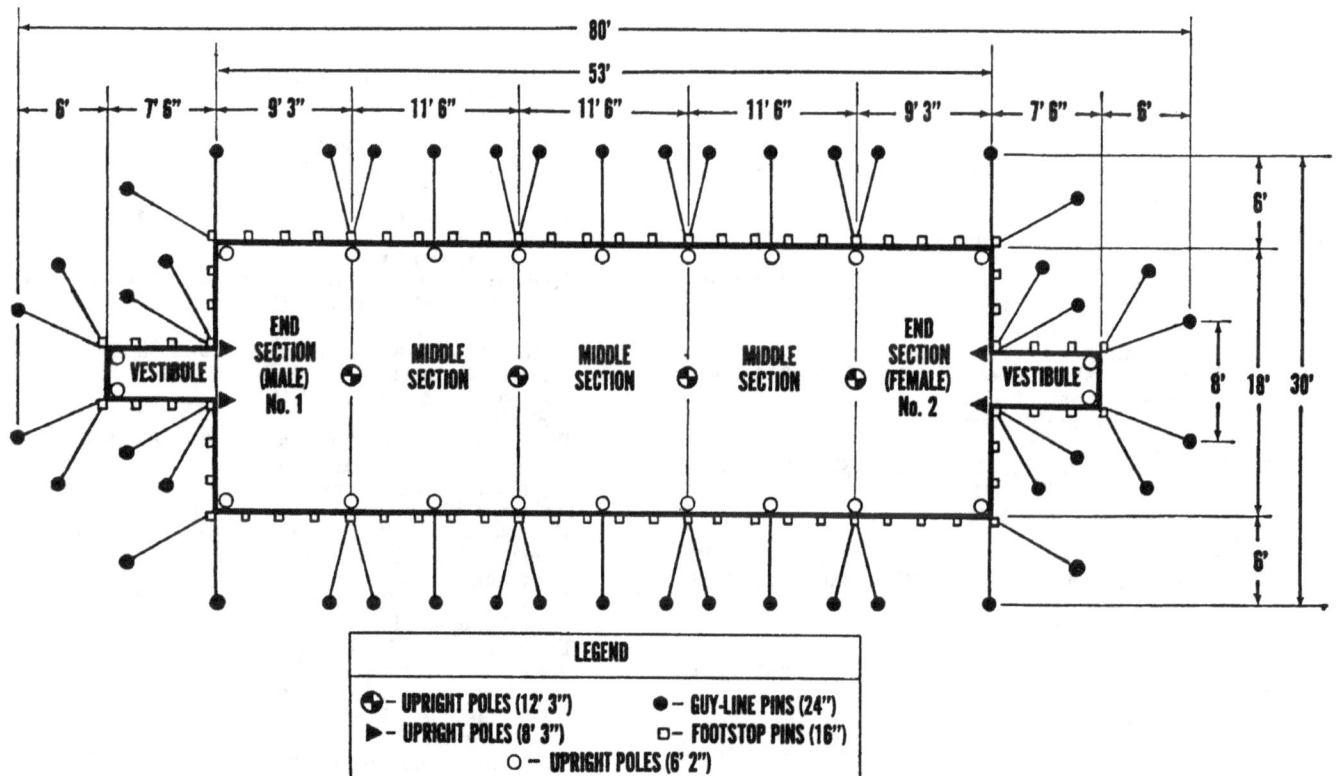

Figure 37. Ground plan of tent, hospital, sectional, FWWMR, OD.

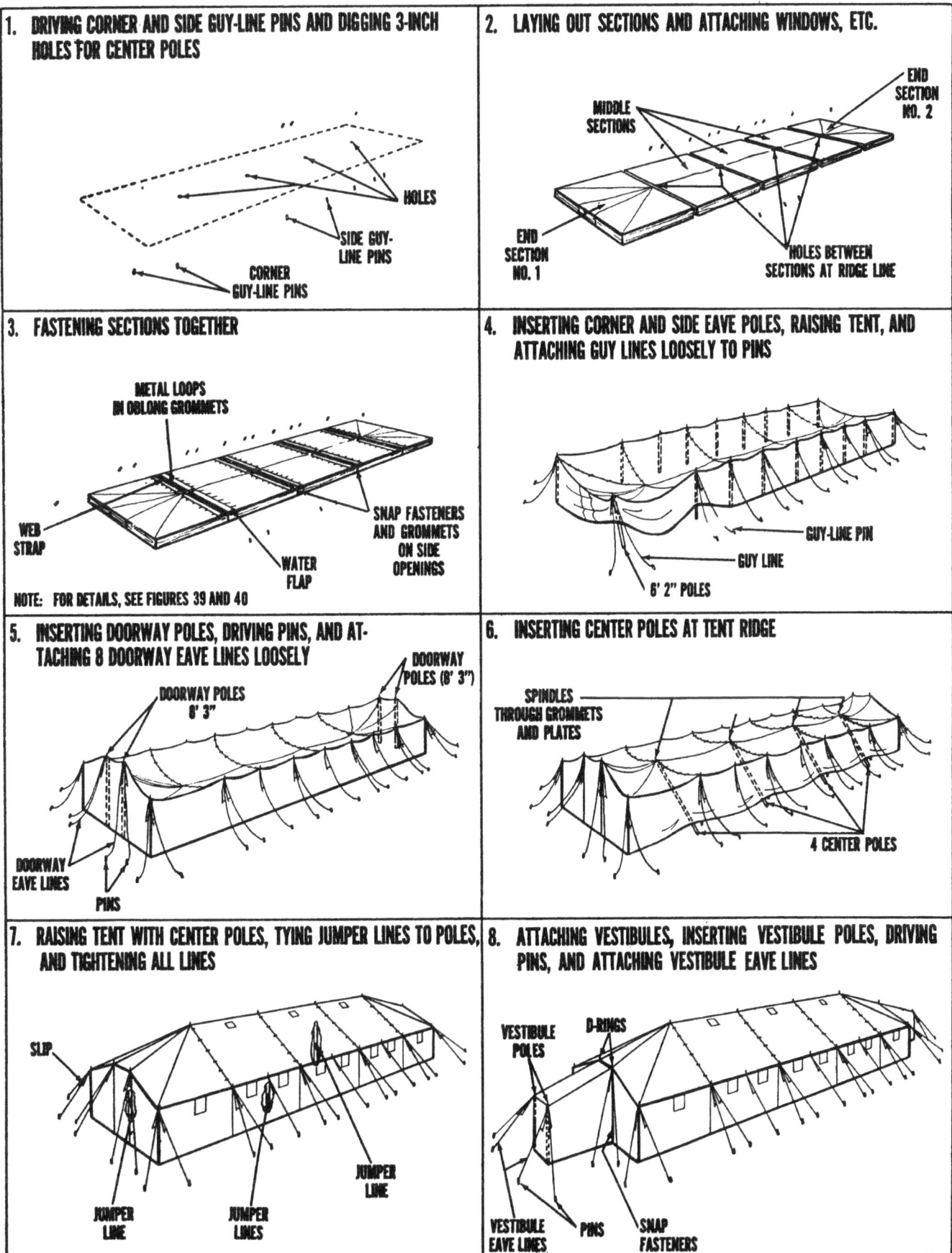

Figure 38. Steps in pitching tent, hospital, sectional, FWWMR, OD.

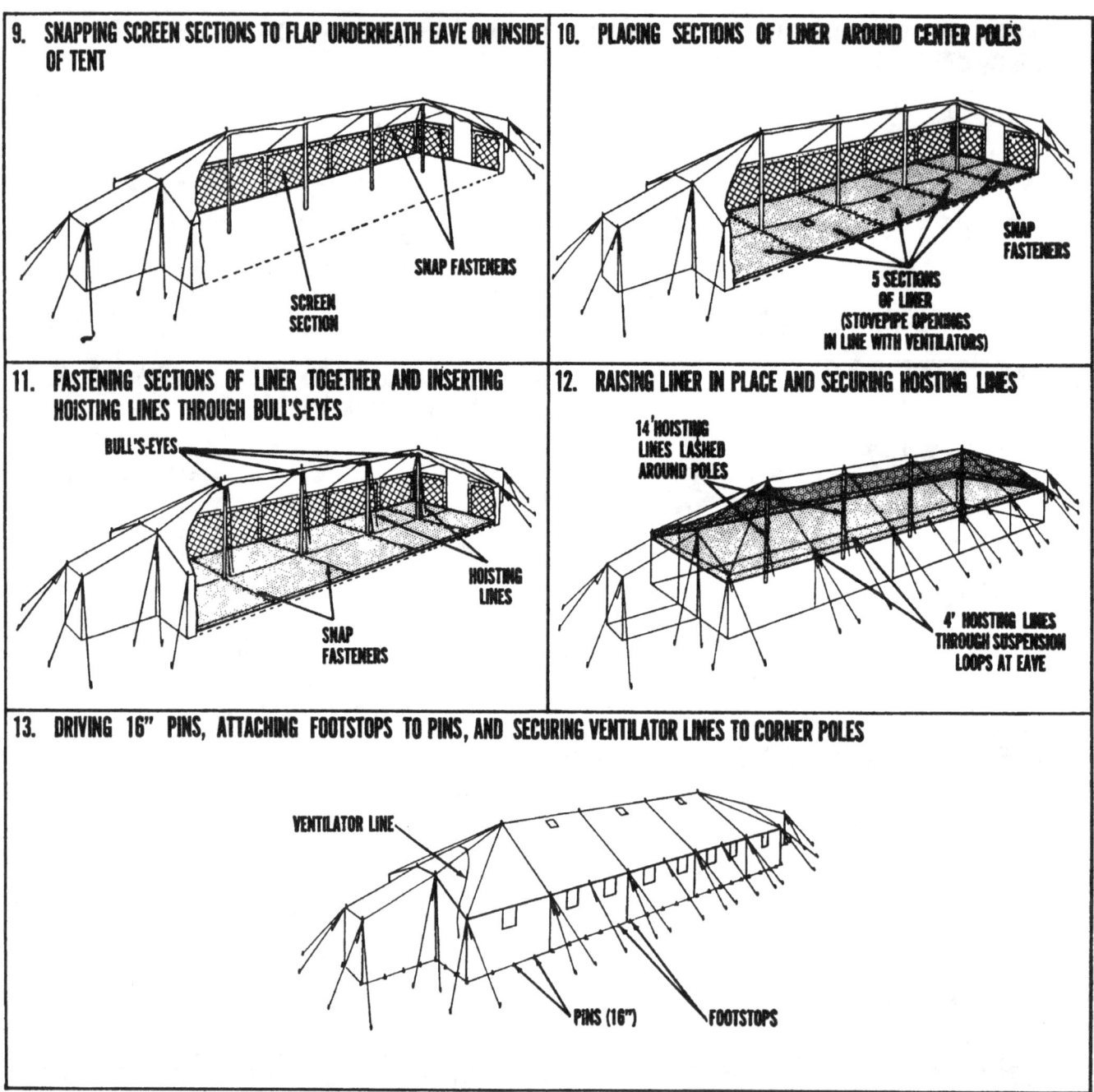

Figure 38—Continued.

(2) Mark locations of center poles and dig holes 3 inches deep. In soft ground it may be necessary to place salvage wood in the bottom of the hole to prevent sinking of center poles.

b. *Laying Out Sections and Attaching Windows, etc.* (2, fig. 38).

(1) Lay out sections of tent on ground in position on tent area, with the No. 1 end section to the left, the No. 2 end section to the right, and the 3 middle sections between the end sections. When laying out middle sections, be sure that ventilators are on the same side of tent.

(2) Attach windows, ventilator sashes, and stovepipe openings, making sure that male studs of snap fasteners face inward.

c. *Fastening Sections Together* (3, fig. 38).

(1) Fasten sections together at side walls by inserting grommets near edge of one section between sockets and studs of snap fasteners near edge of adjoining section and closing snap fasteners (fig. 39).

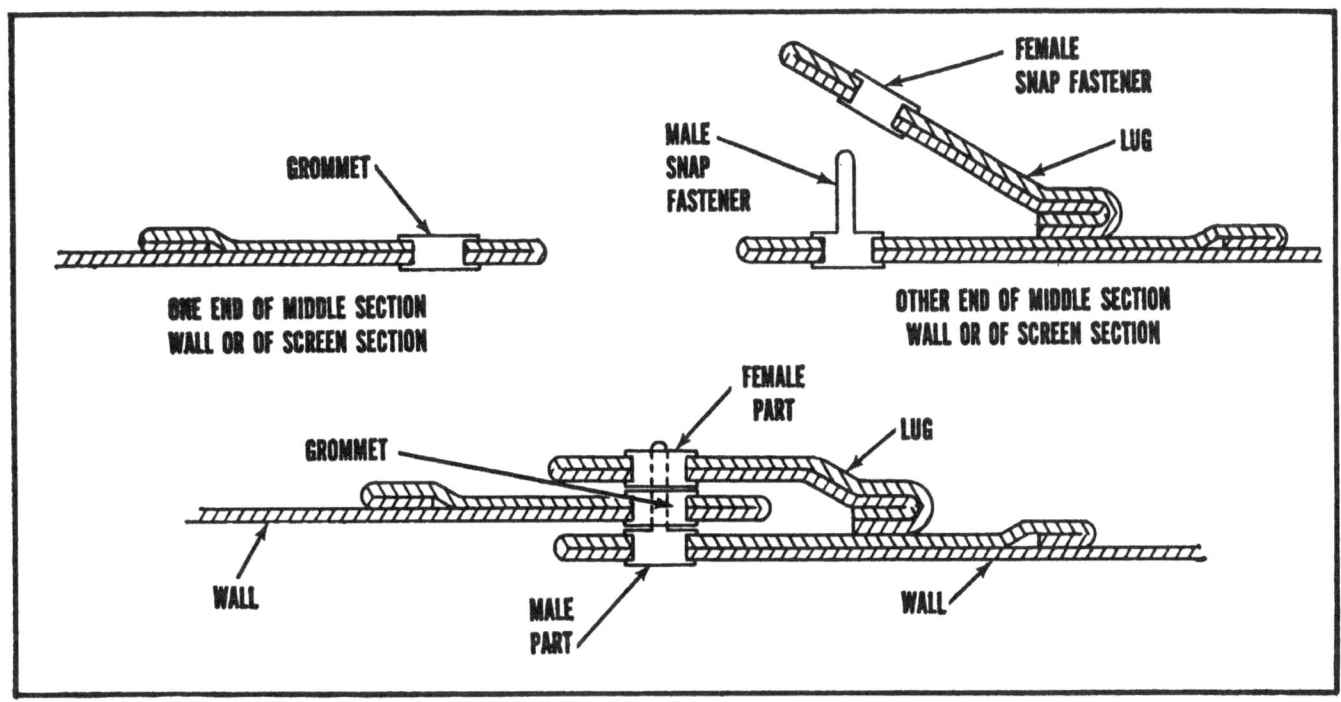

Figure 39. Method of fastening sections together at the side walls or of attaching screen sections (tent, hospital, sectional, FWWMR, OD).

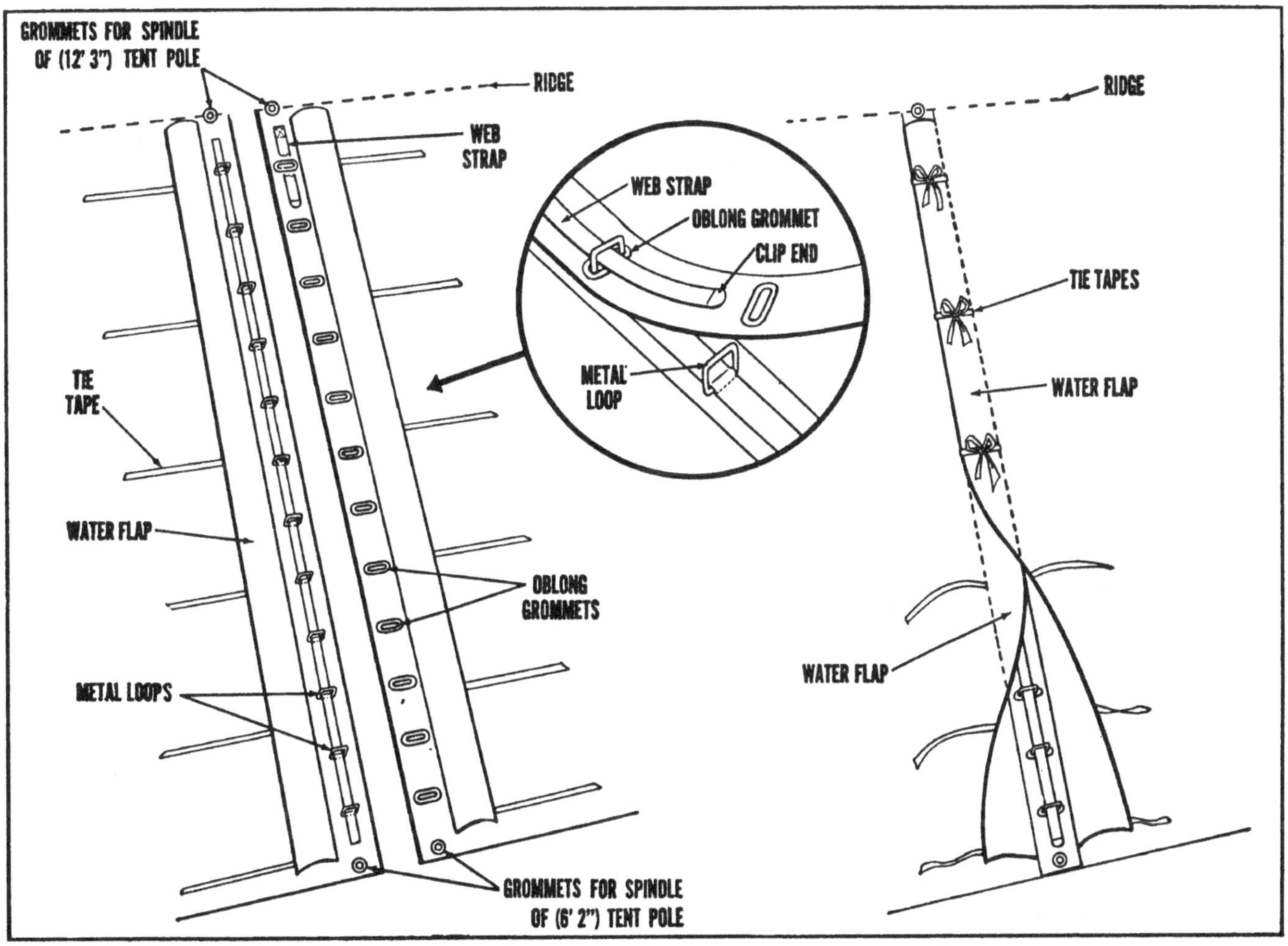

Figure 40. Method of fastening sections together at the top (tent, hospital, sectional, FWWMR, OD).

(2) Fasten sections together at tent top by inserting metal loops near edge of one section through the oblong grommets near edge of adjoining section and lacing web strap through metal loops from ridge downward. Then pull water flap edge of each section over lacing, and tie tie tapes (fig. 40).

d. Inserting Corner and Side Eave Poles, Raising Tent, and Attaching Guy Lines to Pins (4, fig. 33).

    (1) Insert spindles of the 6-foot 2-inch upright corner and side eave poles through plates and grommets at eave line of tent.

    (2) Raise tent at eave line, straighten corner and side eave poles, and attach side eave lines loosely to the 24-inch pins.

e. Inserting Doorway Poles, Driving Pins, and Attaching 8 Doorway Eave Lines Loosely (5, fig. 38).

    (1) Insert spindles of the 8-foot 3-inch doorway poles through plates and grommets over doorway in roof of each end section.

    (2) Drive pins, straighten doorway poles, and attach the 8 doorway eave lines loosely to pins.

f. Inserting Center Poles at Tent Ridge (6, fig. 38). Insert spindles of the 12-foot 3-inch upright center poles through plates and grommets at tent ridge.

g. Raising Tent With Center Poles, Tying Jumper Lines to Poles, and Tightening All Lines (7, fig. 38).

    (1) Raise tent with center poles and place butt of each pole in the 3-inch hole previously prepared in the ground.

    (2) Tie jumper lines to poles and tighten all guy lines so that tent will line up evenly.

h. Attaching Vestibules, Inserting Vestibule Poles, Driving Pins, and Attaching Vestibule Eave Lines (8, fig. 38). Since both vestibules are alike, either one may be used at either end of the tent. At top of vestibule, attach snaps through D-rings located at head of doorway of tent. At sides of vestibule, attach snap fastener studs located along rear edge to snap fastener sockets on tent doorway. Close snap fasteners on vestibule door. When closing snap fasteners, it is advisable to have one man on the inside and one man on the outside. Place 6-foot 2-inch upright poles at vestibule door. Drive 24-inch vestibule eave-line pins and attach vestibule eave lines to pins.

i. Snapping Screen Sections to Flap Underneath Eave on Inside of Tent (9, fig. 38 and fig. 39). When side wall screens are used, work from the inside of tent, inserting grommets in screen sections between sockets and studs of snap fasteners, which are located underneath eaves and at vertical edges of tent sections, and closing snap fasteners.

j. Placing Sections of Liner Around Center Poles (10, fig. 38). Lay sections of liner out on inside of tent around center poles, lining up stovepipe openings of liner with ventilators of tent.

k. Fastening Sections of Liner Together and Inserting Hoisting Lines Through Bull's-Eyes (11, fig. 38).

    (1) Fasten sections of liner together by closing snaps, working from center toward sides.

    (2) Climb up center poles and insert the 14-foot hoisting lines through bull's-eyes in ridge of tent.

l. Raising Liner in Place and Securing Hoisting Lines (12, fig. 38).

    (1) Raise liner in place by pulling hoisting lines. A space of about 12 to 15 inches between the tent and liner is most efficient for insulating purposes.

    (2) Secure the 14-foot hoisting lines around center poles.

    (3) Secure the 4-foot hoisting lines through suspension loops on eave.

m. Driving 16-Inch Pins, Attaching Footstops to Pins, and Securing Ventilator Lines to Corner Poles (13, fig. 38).

    (1) Drive the 16-inch pins and attach footstops to pins. When screens are used and side walls of tent are rolled up and tied, footstops on screens should be attached to the 16-inch pins.

    (2) Secure ventilator lines to corner poles.

43. Striking

a. Remove all footstops and pull out all footstop pins. Bear in mind that as pins, poles, vestibules, and other parts are removed, they should be assembled neatly in front of tent.

b. Loosen door flaps and remove and fold vestibules.

c. Lower, disassemble, and fold liner.

d. Remove and fold screens.

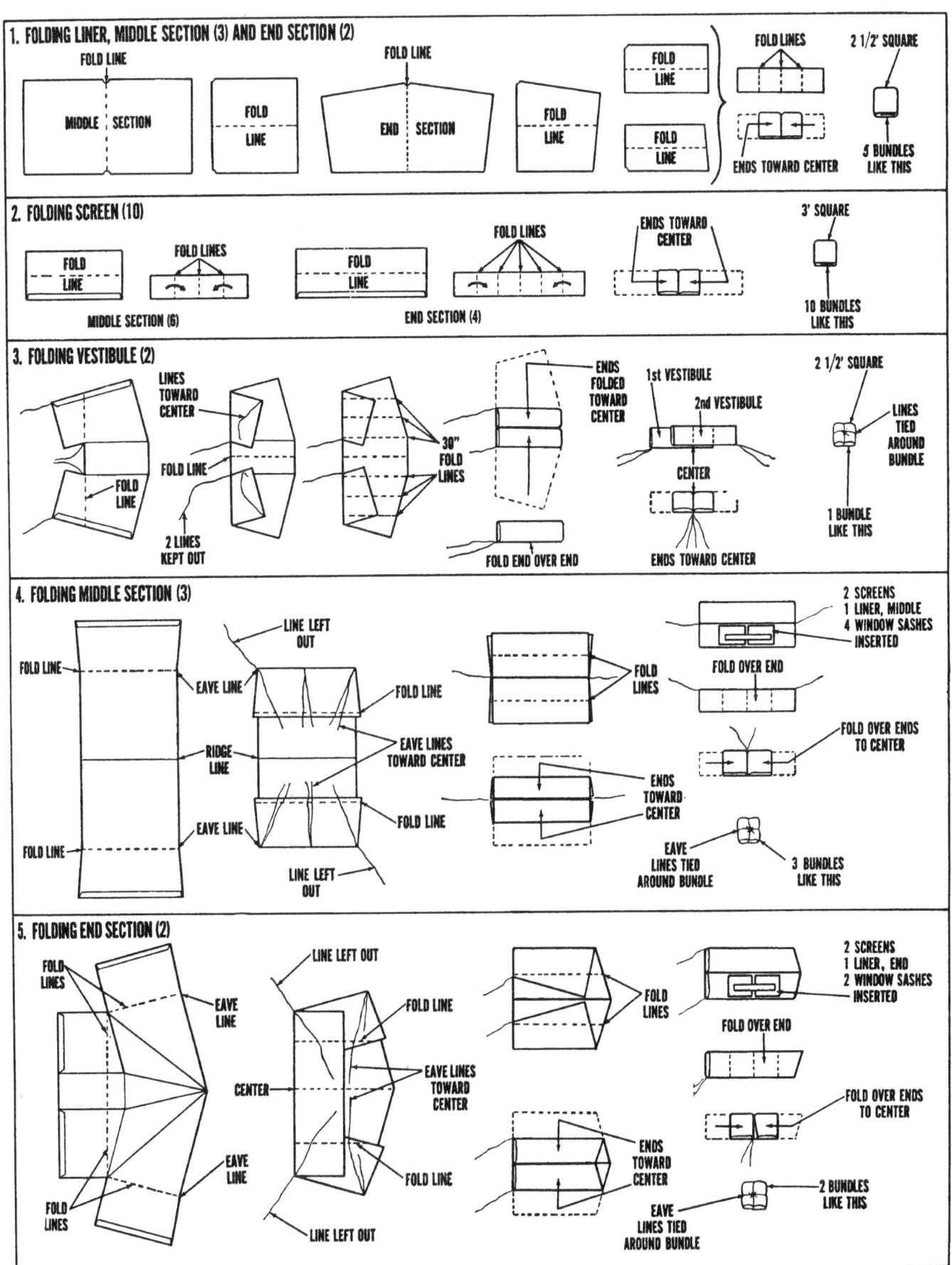

Figure 41. Steps in folding tent, hospital, sectional, FWWMR, OD.

e. Unroll side walls and remove window sashes.

f. Loosen snap fasteners between side wall sections.

g. Unfasten all jumper lines from upright poles.

h. Loosen eave lines and remove 12-foot 3-inch upright center poles.

i. Slip eave lines off pins and pull out eave-line pins.

j. Remove side wall poles and doorway poles.

k. Disassemble tent sections by untying water flaps and unlacing tent top.

44. Folding

a. Folding Liner, Middle Sections and End Sections (1, fig. 41). Fold sections of liner into 5 bundles approximately 2½ feet square. This size permits them to be included in the final folding of the tent section to which each liner belongs. Fold each of the 3 middle section liners and each of the 2 end section liners in half lengthwise, then in half crosswise; then fold ends toward center.

b. Folding Screens (2, fig. 41). Fold each of the 10 screen sections into a bundle approximately 3 feet square. This size permits them to be included in the final folding of the tent section to which each screen belongs. Fold each screen section in half crosswise; then fold ends toward center.

c. Folding Vestibules (3, fig. 41). Spread one vestibule on the ground with weather side down. Fold door flaps over main part of vestibule and coil all but two lines on top. Grasp footstop edge of vestibule and fold ends toward center in 30-inch folds; then fold ends together. Fold the second vestibule in the same manner as the first. The vestibules at this stage of folding should be in 2 bundles approximately 30 inches wide and 7½ feet long. Complete the folding by placing one vestibule on top of the other, staggering them about 30 inches. Fold them together, starting from each end and folding in 30-inch folds to the center. Finish by placing one folded end over the other and tying with the exposed lines.

d. Folding Middle Sections (4, fig. 41). Fold each of the 3 middle sections separately. Pull canvas smooth. Fold side walls over on top of canvas so that fold will be even with eave. Coil all eave lines toward center except two opposite corner lines for tying. Grasping eaves of the section, fold them over so that they meet in center. Fold edges to center once again and then fold one side over the other. After inserting folded liner section, 2 folded screen sections, and 4 window sashes, complete by folding from each end to center and placing one folded end over the other. Using the 2 loose eave lines, secure section by crossing lines at right angles about bundle. Tie with a slipknot.

e. Folding End Sections (5, fig. 41). Fold each of the 2 end sections separately. Smooth canvas and fold end wall toward center even with eave. Complete the folding in the same manner as the middle section (*d* above). When finished, each end section bundle should contain 1 folded end section, 2 folded end section screens, 1 folded end section liner, and 2 window sashes.

Section VIII. KITCHEN TENT

45. Use

The tent, kitchen, flyproof, M-1948, FWWMR, OD, complete with pins and poles (fig. 42), is a screened shelter used for cooking and serving food where flies and other insects are numerous.

46. Description

The tent is an A-shaped, square-ended, rectangular tent. The back or field range section of the tent forms a stack, elevated in turret fashion, 3 feet higher than the front or service section of the tent, in order to accommodate the field ranges. Both sections have a similar contour, sloping gently to each side of a central ridge. The side and front walls of the tent may be guyed out, forming awnings on the side and front. A wall screen which snaps to the tent provides an insect-proof closure on sides and front when the walls are raised. The tent may be completely blacked out.

a. Dimensions. The tent is approximately 12 feet wide and 18 feet long. The ridge height of the tent is 9 feet at the service section and 12 feet at the stack section. The roof slope is approximately 6 inches per running foot, or slightly less than 3 feet from ridge to side wall. The end and side walls are vertical, the side walls being 6 feet high.

b. Weight and Cubage. The tent and screen weigh 202.5 pounds, the poles 171.5 pounds, and

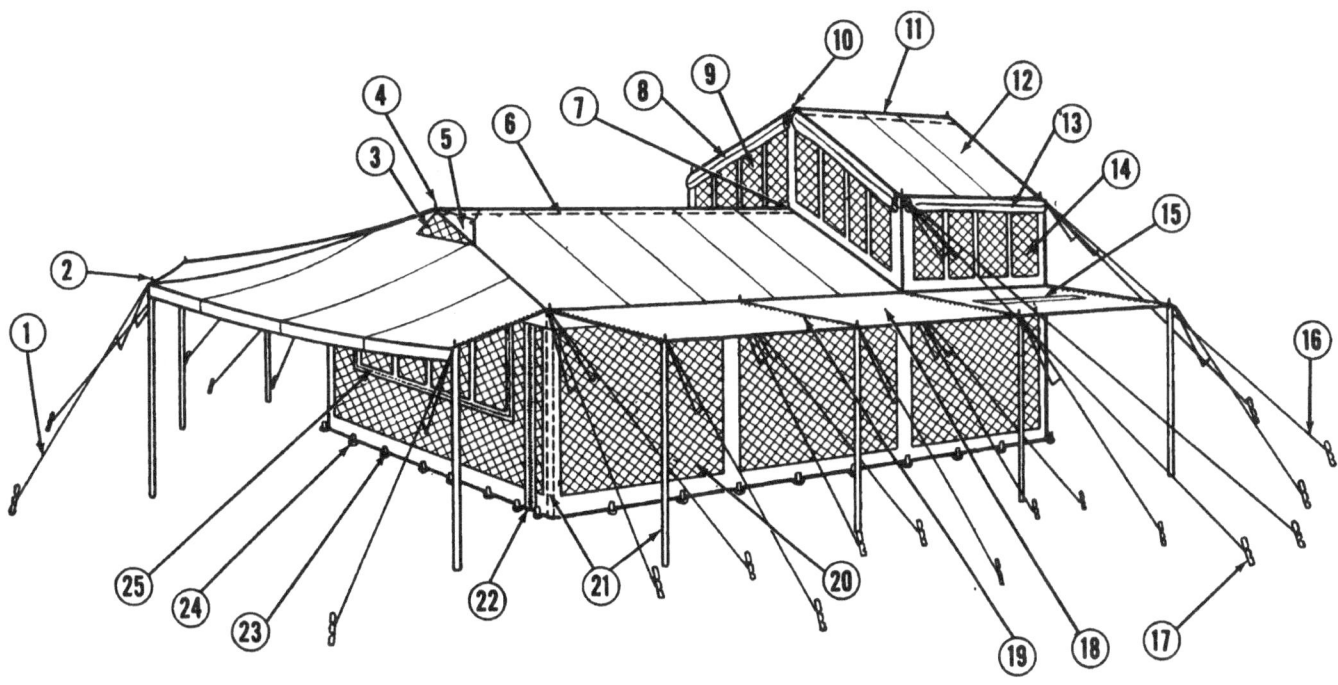

Reference No.	Part	Quantity	Federal Stock No.
1	Line, tent, manila, one end sewed, one end w/eye, 5/16" dia., 11' long (eave line)	23	8340-252-2268
2	Pole, tent, upright, solid, 7'	1	8340-188-8407
3	Screen, ventilator, front	2	
4	Pole, tent, upright, solid, 9'	5	8340-188-8410
5	Flap, ventilator, front	2	
6	Pole, tent, ridge, jointed, metal sleeve 11' 10"	1	8340-188-8396
7	Spindle of 11' 10" ridge pole	1	8340-223-7894
8	Flap, front stack ventilator	2	
9	Screen, front stack ventilator	1	
10	Pole, tent, upright, jointed, metal sleeve 12' 3"	2	8340-188-8411
11	Pole, tent, ridge, solid, 5' 11¼"	1	8340-188-8392
12	Stack roof	1	
13	Flap, stack side	2	
14	Screen, stack side	2	
15	Ventilator opening, side wall	2	
16	Line, tent, manila, one end sewed, one end w/eye, 5/16" dia., 19' long (guy line)	8	8340-252-2273
17	Pin, tent, wood, 24"	31	8340-261-9751
18	Side wall awning section	6	
19	Slide fastener, awning	8	
20	Tent screen, wall, kitchen, flyproof, M-1948, plastic, FWWMR, OD	1	8340-254-8997
21	Pole, tent, upright, solid, 6' 2"	16	8340-188-8406
22	Slide fastener, entrance opening	1	
23	Pin, tent, wood, 16"	32	8340-261-9750
24	Line, tent, cotton, footstop, ¼" dia., 19" long (footstop line) (26 on screen)	67	8340-252-2299
25	Slide fastener, service window opening	3	
	Line, tent, cotton, 9', unfinished, ⅛" dia. (side flap hoisting and hood line)	18	8340-252-2301
	Line, tent, manila, one end sewed, ¼" dia., 6' long (jumper line)	10	8340-252-2286
	Line, tent, manila, one end sewed, ¼" dia., 10' 6" long (jumper line)	4	8340-252-2282
	Line, tent, cotton, 14' long, unfinished, ⅛" dia. (flap tie line and center flap hoisting line)	4	8340-252-2303
	Line, tent, cotton, 3' 4" long, both ends sewed, ¼" dia. (screen wall tie line)	1	8340-252-2305
	Slip, tent line, steel	31	8340-223-8094
	Cover, tent, FWWMR, kitchen, flyproof, M-1948	11	8340-241-8436
	Line, tent, manila, one end sewed, one end w/eye, 5/16" dia., 13' long (cover tie line)	2	8340-252-2271

Figure 42. Tent, kitchen, flyproof, M-1948, FWWMR, OD, complete with pins and poles, Federal Stock No. 8340-262-3687.

the pins 45 pounds. The tent and screen have a cubage in storage of 14.2 cubic feet, the poles a cubage of 9 cubic feet, and the pins a cubage of 3 cubic feet.

c. *Floorspace.* The floorspace is approximately 216 square feet. The ground area of the stack section provides space for three M-1937 field ranges. The service area provides space for a large kitchen table and utensils.

d. *Material.* The tent is made of 12.29-ounce duck, FWWMR.

e. *Support.* The tent is supported by 13 upright poles and 2 ridge poles. When the sides and front are guyed out to form awnings, 11 additional upright poles are required.

f. Entrances. The vertical side walls of the tent are equipped with 8 slide fasteners. Entrance to the tent may be gained by opening any one of the fasteners. However, when the screen wall is attached, entrance may be gained only by opening the one slide fastener in the corner near the service window.

g. Ventilation.

(1) The 4 elevated sides of the stack section are equipped with air-permeable screening. This induces a draft, so that heat from the field ranges is taken away through the screening.

(2) When conditions are favorable, slide fasteners can be released and the side walls and front end of the tent lifted from the bottom to provide increased ventilation. The side and front walls can be guyed out, with the bottom seams supported by tentpoles to form awnings.

(3) When the side walls and front end of the tent are lowered and the slide fasteners closed to bar passage of light and to provide safe blackout operation conditions, adequate ventilation may be obtained by adjusting the ventilator flaps on the sides and the rear of the stack section near the base and on the front of the service section near the ridge pole.

h. Screens.

(1) A detachable screen, made of 2.9-ounce type II nylon cloth, snaps to the tent and may be used as an insectproof closure on the sides and front of the tent. The screen is fastened at the top of both end and side tentpoles and drapes vertically to the ground, where it is attached to 16-inch tent pins. The screen has a service window in the front section which may be opened by slide fasteners and rolled up. The front end and side walls of the tent may be raised and guyed out to form awnings, in which case the screen wall offers insect protection.

(2) There are built-in screens on the front and sides of the stack section near the top. Flaps operated by hoisting lines may be used to cover or uncover the screens. There are also built-in screens on the sides and rear of the stack section near the base and on the front of the service section near the ridge pole.

i. Cover. The tent is provided with a cover for use when in storage or when being transported.

47. Ground Plan

Before pitching the tent, study the ground plan carefully (fig. 43).

48. Pitching

The tent can be pitched by 5 men in approximately 60 minutes. When conditions permit, the tent should be pitched away from natural elevations or tall equipment that might obstruct a draft through the tent stack.

a. Spreading Tent Out With Side Poles and Pins in Position (1, fig. 44). Spread tent out according to ground plan with the four 9-foot and the six 6-foot 2-inch side poles and twenty 24-inch pins in proper position.

b. Driving Pins on One Side of Tent, Raising Side With Poles Pointing Slightly Toward Inside, and Attaching Guy Lines (2, fig. 44).

(1) Drive ten 24-inch pins on one side of tent site, according to ground plan.

(2) Insert spindles of two 9-foot upright poles through grommets in eave at one side of stack section of tent, raise stack section side, and attach guy lines to pins.

(3) Insert spindles of three 6-foot 2-inch upright poles through grommets in eave at one side of service section of tent, raise service section side, and attach guy lines to pins.

c. Raising Other Side, Straightening Poles, Closing Slide Fasteners, Driving 16-inch Pins, and Attaching Footstops to Pins (3, fig. 44).

(1) Raise the other side of tent in the same manner as the first side.

(2) Straighten poles, close slide fasteners, drive 16-inch pins, and attach footstops to pins.

d. Fastening Short Ridge Pole to 12-Foot 3-Inch Upright Center Pole and Raising (4, fig. 44). Insert spindles of the two 12-foot 3-inch upright center poles through holes at ends of short ridge pole, raise poles, and insert spindles of the 12-foot 3-inch upright center poles into grommets in stack ridge. Make sure that the 12-foot 3-inch upright poles are at the front and rear center of stack section 6 feet from each side and that they are perpendicular.

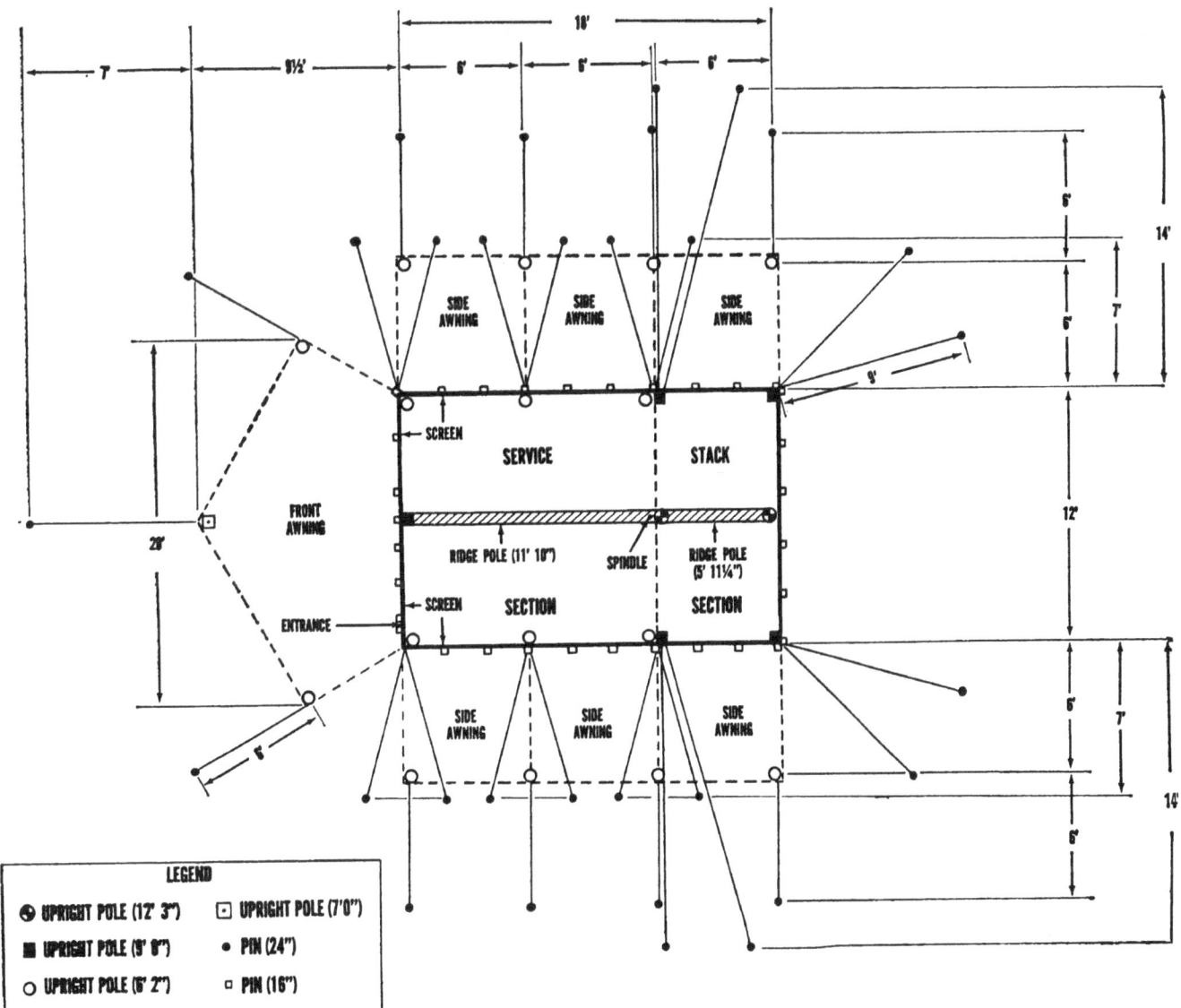

Figure 43. Ground plan of tent, kitchen, flyproof, M-1948, FWWMR, OD.

e. Raising Long Ridge Pole, Fastening It to 9-Foot Upright Front Pole, and Securing It to 12-Foot 3-Inch Upright Center Pole (5, fig. 44).

(1) With one man at each end of the long ridge pole, raise pole to a position where a third man can insert the spindle of the 9-foot upright front pole through the hole in the front end of the ridge pole and into the grommet in the service ridge of the tent; then set upright pole in place perpendicularly as indicated on ground plan.

(2) Fasten connector end of long ridge pole to the 12-foot 3-inch upright front center pole (fig. 45) about 3 feet from the top of the stack so that the long ridge pole is level with the ground; this is done by placing connector of ridge pole around upright pole, swinging swivel plate into position on one side of upright pole, and tightening nut. Attach jumper line at front stack ridge around short ridge pole with a half hitch and secure it to metal loop of connector with a round turn and 2 half hitches. Insert spindle of connector through grommet at ridge at rear end of service section of tent.

f. Extending Side and Front Walls to Form Awnings and Spreading Screen (6, fig. 44).

(1) Unfasten slide fasteners at front and rear corners of side walls and detach footstops from pins. Extend side walls outward with eight 6-foot 2-inch poles and front wall with two 6-foot 2-inch poles

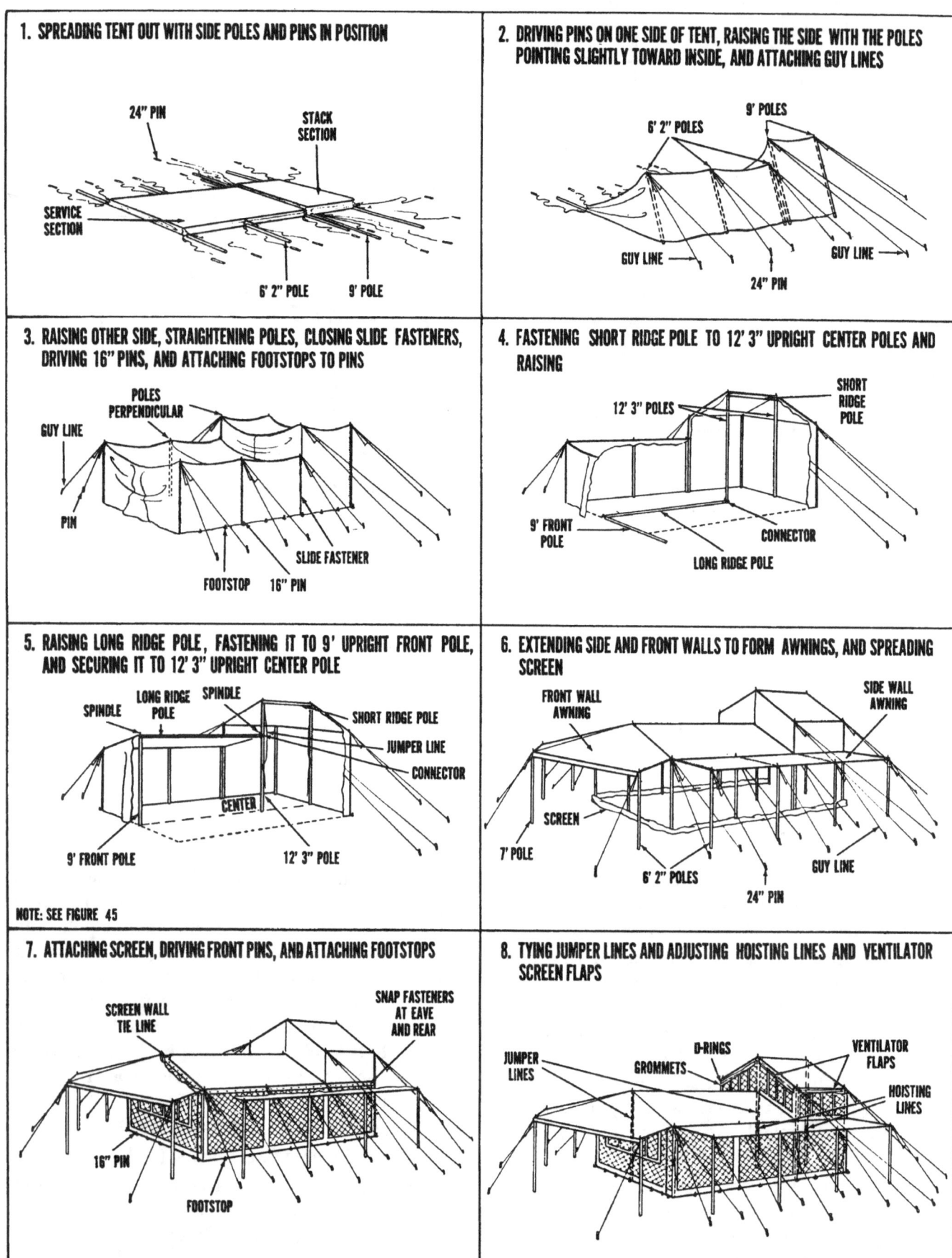

Figure 44. Steps in pitching tent, kitchen, flyproof, M-1948, FWWMR, OD.

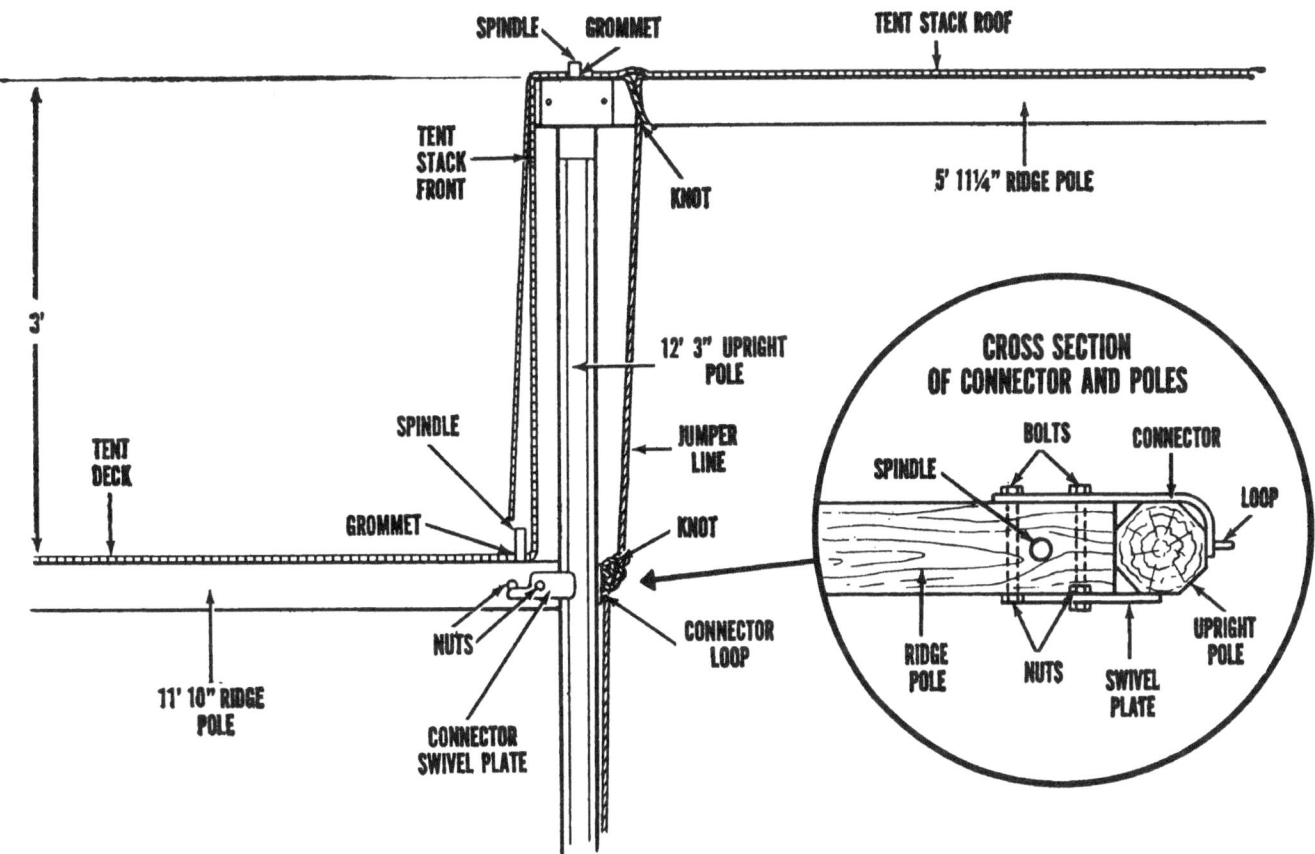

Figure 45. Fastening ridge poles to 12-foot 3-inch upright center pole (tent, kitchen, flyproof, M–1948, FWWMR, OD).

and one 7-foot pole to form awnings. Drive pins and attach guy lines, according to ground plan. The slide fasteners on the side wall awnings and at the front end of the stack may be unfastened, and the long guy lines from the 9-foot front side stack poles may go through the openings.

(2) Spread out screen around outside of side and front poles at base of tent.

g. Attaching Screen, Driving Front Pins, and Attaching Footstops (7, fig. 44).

(1) Hang screen to tent by fastening snap fasteners at eave and rear corners. Remove tops of side wall poles from eave grommets, insert spindles of poles into tabs in screen, and replace poles. Tie front peak of screen to ridge pole with screen wall tie line.

(2) Drive the remaining 16-inch pins and attach footstops to pins in front of screen; attach footstops on sides of screen to 16-inch pins already driven.

h. Tying Jumper Lines and Adjusting Hoisting Lines and Ventilator Screen Flaps (8, fig. 44).

(1) Tie all jumper lines to eave and center poles.

(2) Adjust hoisting lines which go through D-rings on stack ventilator flaps and grommets on screen panel. Raise flaps and tie hoisting lines around stack upright poles.

i. Closing Tent for Blackout (fig. 46). Remove awning poles. Drop awnings and close slide fasteners. Place footstops over the same pins as are used for the screens. Close top stack section ventilator flaps. Make sure that top front service

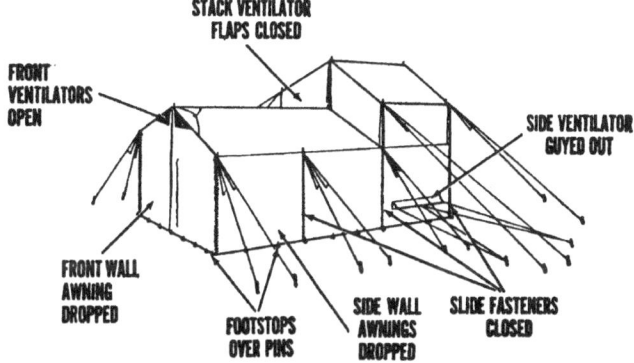

Figure 46. Closing tent for blackout (tent, kitchen, flyproof, M–1948, FWWMR, OD).

section ventilator flaps are open. Tie lines of bottom stack section ventilator hoods in sides and rear to pins to assure a draft through the tent.

49. Striking

a. Unfasten slide fasteners along tent side walls.

b. Release footstops and remove the 16-inch pins.

c. Raise awnings temporarily, using the 6-foot 2-inch poles.

d. Incline tentpoles supporting screen, remove screen from spindles of poles, and reset poles.

e. Close all ventilator flaps.

f. Remove awning poles and drop awnings.

g. Remove the 3 center poles and their connecting ridge poles.

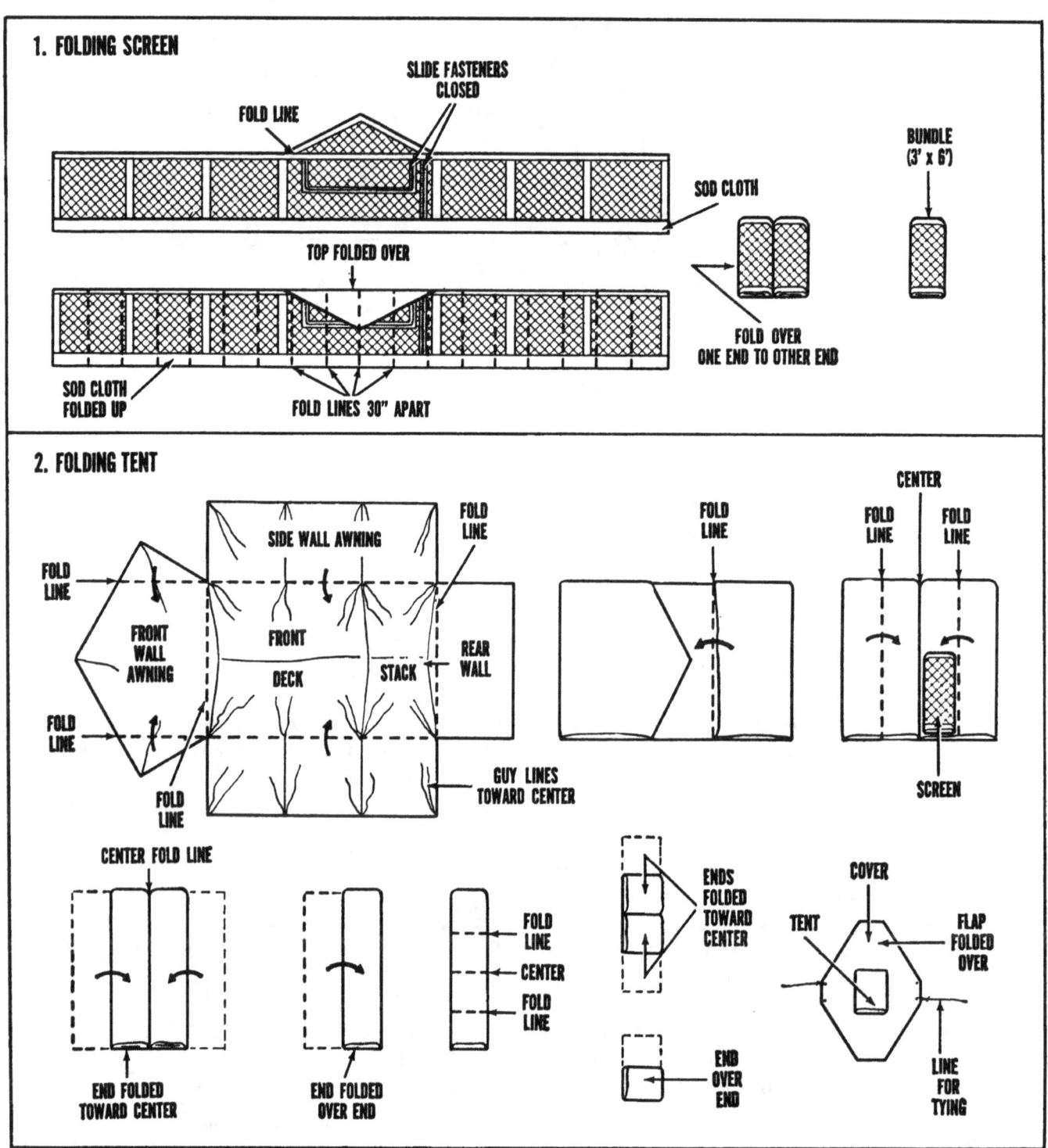

Figure 47. Steps in folding tent, kitchen, flyproof, M–1948, FWWMR, OD.

h. Remove the 6-foot 2-inch service section side wall poles.

i. Remove the 9-foot stack section side wall poles.

j. Remove the 24-inch pins from ground

50. Folding

a. Folding Screen (1, fig. 47). Spread screen flat on ground and close slide fasteners. Fold in sod cloth and triangular part at top of front section to form a straight line. Fold ends toward center in 30-inch folds. Fold one end over the other end, making a 3- by 6-foot bundle.

b. Folding Tent (2, fig. 47).

(1) Spread tent flat on ground, arranging as neatly as possible. Throw all guy and eave lines toward center. Fold side and end walls toward center.

(2) Grasp corners of rear wall and stack; fold over on front deck. The tent is now a 12-foot square. Establish a center line and place folded screen to right of center line.

(3) Fold ends toward center and end over end. Establish a center line. Fold ends toward middle and end over end. The tent is now approximately a 3- by 3-foot bundle.

(4) Place in cover. Fold long flaps over first, then the shorter ones. Secure bundle with lines tied through grommets and around bundle.

Section IX. LATRINE SCREEN

51. Use

The screen, latrine, FWWMR, complete with cover, pins, and poles (fig. 48), is issued to units in the field for use as an outdoor latrine. It may also be used by the Identification, Effects, and Record Section of Quartermaster Graves Registration Company to conceal remains from view until identification procedures and preparation of remains for burial have been completed.

52. Description

a. Dimensions. The screen is a canvas panel, 55 feet long and 5 feet 3 inches wide. When erected, the screen is rectangular, 18 feet long, 9 feet wide at one end, and 7 feet wide at the other end. The difference in width is for the purpose of forming a 2-foot entrance on one side of the screen. The entrance side consists of a 12-foot section and a 9-foot section, which overlap by approximately 3 feet to give depth to the entrance.

b. Floorspace. The floorspace of the screen, when erected, is approximately 144 square feet.

c. Material. The screen is made of 9.68-ounce duck.

d. Cover. The screen is provided with a cover for use when in storage or when being transported.

53. Ground Plan

Before pitching screen, study ground plan carefully (fig. 49).

54. Pitching

The screen can be pitched by 6 men in approximately 20 minutes.

a. Spreading Out Screen, Driving Pin at Lower Inside Corner of Entrance, and Attaching Short Tie Line (1, fig. 50).

(1) Unfold screen and lay it out on ground, following ground plan. Place end having both a long and a short tie line, as differing from the end having two short tie lines, at inside corner of entrance.

(2) Drive a tent pin at inside corner of entrance, following ground plan. Attach short tie line on bottom of screen to this pin.

b. Connecting Upright Poles to Ridge Poles (2, fig. 50). Connect upright poles to ridge poles as follows:

(1) Two upright entrance poles and one upright side pole to a 9-foot ridge pole.

(2) Two upright side poles to one 7-foot ridge pole.

(3) Two upright side poles to one 9-foot ridge pole (before assembling, loop the 11-foot tie line over ridge pole).

c. Raising Three Center Poles in Position Inside Screen, Driving Pins, and Attaching Guy Lines (3, fig. 50). Raise the 3 center upright poles connected to the 9-foot ridge pole to a vertical position on inside of screen near center. Hold upright

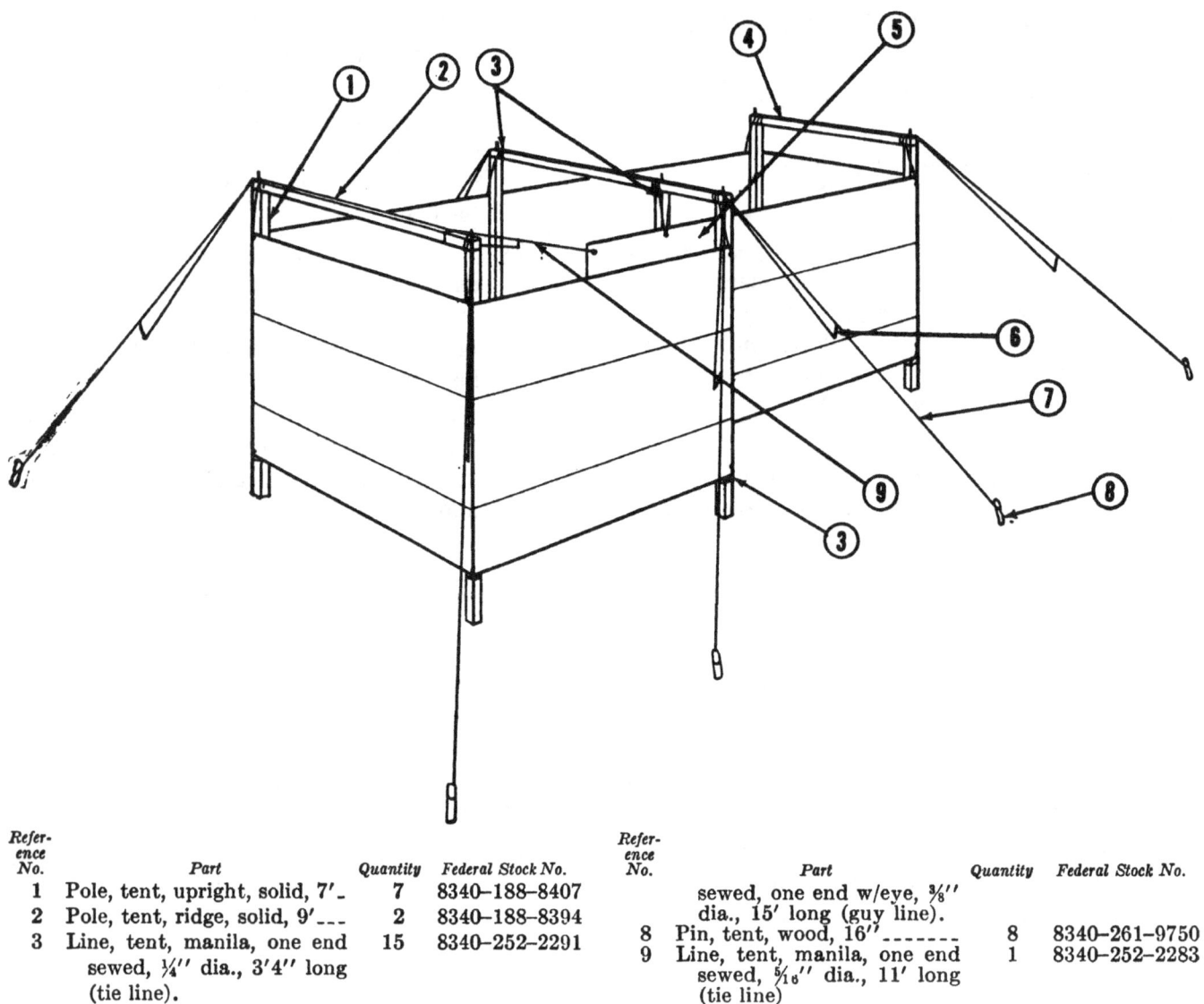

Reference No.	Part	Quantity	Federal Stock No.
1	Pole, tent, upright, solid, 7'.	7	8340-188-8407
2	Pole, tent, ridge, solid, 9'.	2	8340-188-8394
3	Line, tent, manila, one end sewed, ¼" dia., 3'4" long (tie line).	15	8340-252-2291
4	Pole, tent, ridge, solid, 7'.	1	8340-188-8393
5	Entrance	1	
6	Slip, tent line, steel.	8	8340-223-8094
7	Line, tent, manila, one end sewed, one end w/eye, ⅜" dia., 15' long (guy line).	8	8340-252-2270
8	Pin, tent, wood, 16".	8	8340-261-9750
9	Line, tent, manila, one end sewed, ⁵⁄₁₆" dia., 11' long (tie line)	1	8340-252-2283
	Cover, screen, latrine.	1	
	Line, tent, manila, one end sewed, one end w/eye, ¼" dia., 8' long (cover tie line)	2	8340-252-2269

Figure 48. Screen, latrine, FWWMR, complete with cover, pins, and poles, Federal Stock No. 8340-237-8752.

poles in position and drive in 2 center guy-line pins 6 feet from outside door upright pole, according to ground plan. Drive a third guy-line pin 6 feet from upright pole on opposite side of screen and in line with the 3 upright poles. Place guy lines over these pins and over spindles of upright poles.

d. Raising Screen at Center and Tying to Ridge Pole (4, fig. 50). Raise screen at center and tie short tie lines to center ridge pole so that screen is 6 inches off ground at bottom.

e. Raising 2 End Poles in Position, Driving Pins, Attaching Guy Lines, and Tying Screen to Ridge Pole (5, fig. 50).

(1) At narrow end of screen, raise the 2 end upright poles, connected to the 7-foot ridge pole to a vertical position.

(2) Hold upright poles in position and drive in an end guy-line pin 6 feet from each upright pole, according to ground plan. Place guy lines over these pins and over spindles of upright poles.

(3) Raise screen at end and tie short tie lines to ridge pole so that screen is 6 inches off ground at bottom.

f. Raising the Other Two End Poles in Position, Driving Pins, Attaching Guy Lines, and Tying Screen to Ridge Pole (6, fig. 50).

(1) At wide end of screen, raise the 2 end

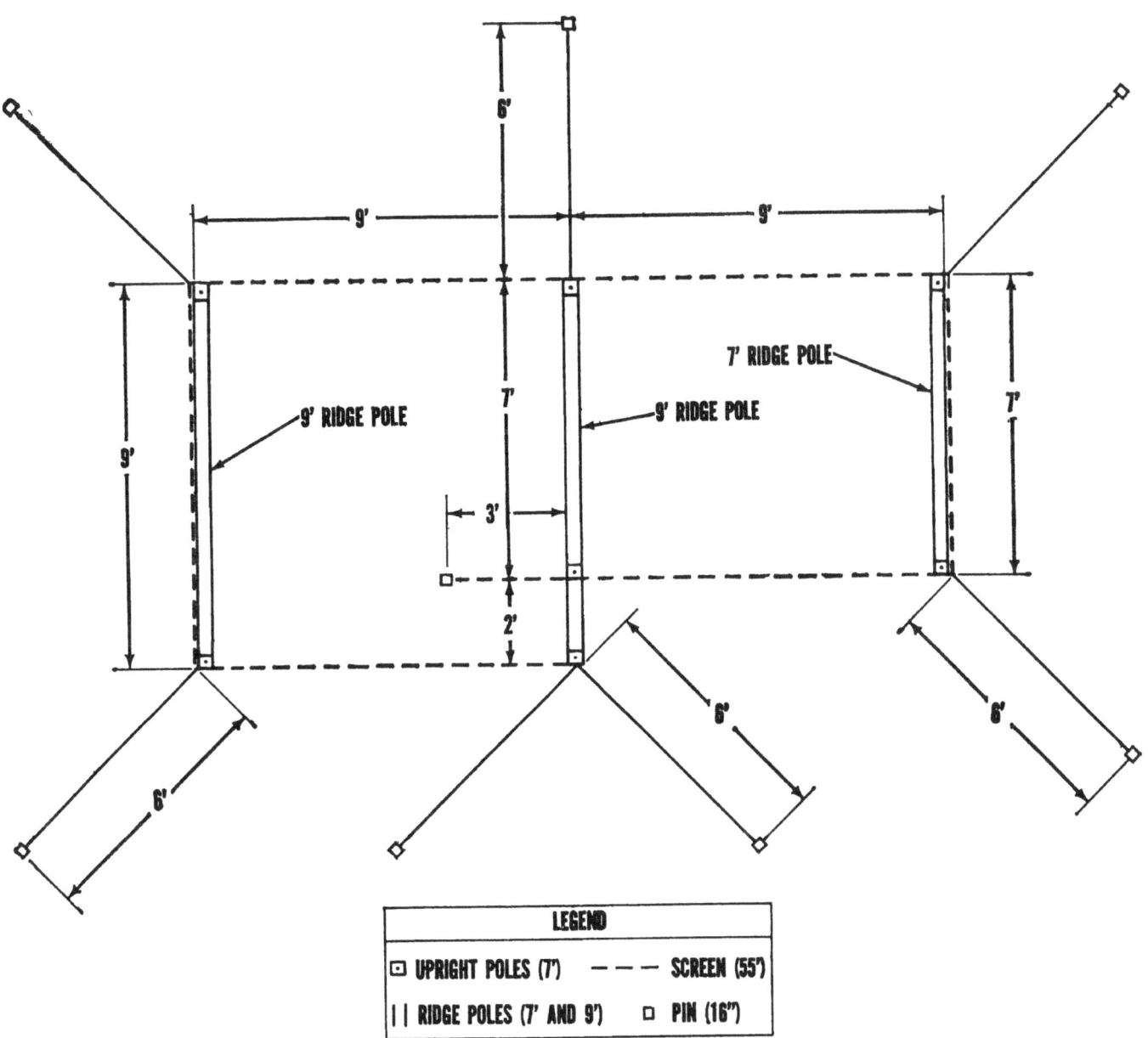

Figure 49. Ground plan of screen, latrine, FWWMR, OD.

upright poles connected to the 9-foot ridge pole to a vertical position.

(2) Hold upright poles in position and drive in an end guy-line pin 6 feet from each upright pole, according to ground plan. Place guy lines over these pins and over spindles of upright poles.

(3) Raise screen at end and tie short tie lines to ridge poles so that screen is 6 inches off ground at bottom. Adjust and tighten the long (11-foot) tie line from top of inside entrance of screen to the end 9-foot ridge pole so that screen is 6 inches off ground at bottom.

g. Placing End of Screen Around Outside Entrance Pole and Tying to Center Ridge Pole (7, fig. 50). Place end of screen around outside entrance upright pole and tie short tie line at end of screen to center ridge pole so that screen is 6 inches off ground at bottom.

h. Tying Bottom of Screen to Poles and Adjusting Guy Lines (8, fig. 50). Tie all short tie lines at lower edge of screen to upright poles. Adjust and tighten all guy lines.

55. Striking

a. Untie all tie lines on lower edge of screen from upright poles.

b. Untie outside of entrance from center ridge pole.

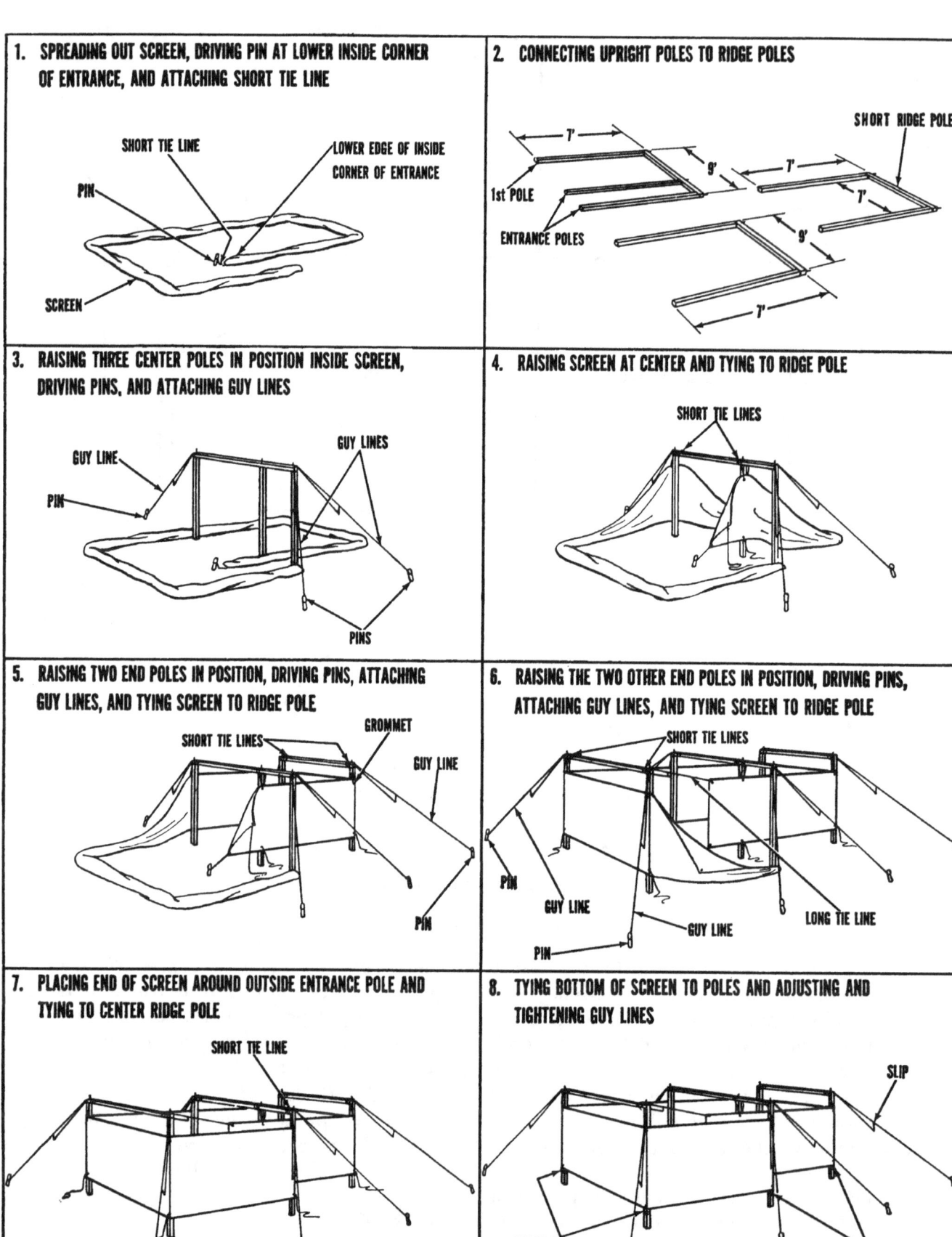

Figure 50. Steps in pitching screen, latrine, FWWMR.

c. Untie screen at wide end and drop it to the ground. Detach guy lines at wide end, and disassemble ridge pole and two upright poles and place them in a pile to one side.

d. Untie narrow end of screen. Detach guy lines and place ridge pole and upright poles with poles previously placed to one side.

e. Follow same procedure with center ridge pole. Collect the 8 tent pins and 7 guy lines and place them near ridge poles and upright poles.

56. Folding

a. Place screen flat on ground and smooth it out.

b. Make two folds. To make the first fold, pull one end over the other. Repeat this step, placing folded edge even with the two ends of the screen. Be sure to smooth out canvas after each fold. The screen can be controlled better if the first two folds are made into the wind.

c. Continue folding. Fold either top or bottom edge over one-third of width of screen. Then fold other edge completely over first fold. Put tie lines and guy lines inside folds at one end of screen.

d. Form final bundle by making a 2-foot fold from each end of screen toward center. Repeat this step twice, leaving the two folded sections 2 or 3 inches apart. Place one folded section over the other. Place bundle on cover, folding edges of cover in. Tie with cover tie lines.

Section X. MAINTENANCE SHELTER TENT

57. Use

The tent, maintenance shelter, FWWMR, OD, complete with frame and pins (fig. 51), is used for the repair of small tanks and trucks.

58. Description

The tent is an A-shaped, rectangular, square-end tent. The tent is erected over a box steel frame supplied for the purpose (fig. 52).

a. *Dimensions.* The tent is 18 feet 2 inches wide, 26 feet 9½ inches long, and 13 feet 7¾ inches high. The wall height is 5 feet 6 inches, giving a pitch of 8 feet 1¾ inches.

b. *Weight and Cubage.* The tent weighs 500 pounds, and the frame weighs 755 pounds. The tent has a cubage in storage of 26.3 cubic feet, and the frame has a cubage of 58 cubic feet.

c. *Floorspace.* The floorspace is 486.8 square feet.

d. *Materials.* The top, side walls, and all reinforcements are made of 12.29-ounce duck. The sod cloth, which is 29½ inches wide, is made of 9.85-ounce duck. There are 6 ground cloths, measuring 4 feet by 12 feet each, provided with each tent to form a floor. The ground cloths are made of No. 6 duck.

e. *Roof Opening.* A section of the roof of the tent may be lowered by slide fasteners operated by draw lines to give an opening approximately 10 by 10 feet through which heavy equipment may be handled by a crane outside the tent.

f. *Ventilation.* The tent is equipped with 2 canvas ventilators, one at each end of the tent near the ridge.

g. *Heating.* The tent may be heated by an external gasoline tent heater. Four heater ducts are provided. Two ducts are located at the rear bottom corner of each of the 2 sides of the tent. When the ducts are not in use, they may be covered by canvas heater duct flaps.

h. *Cover.* The tent is provided with a cover for use when in storage or when being transported.

59. Pitching

The tent, including the steel frame, can be pitched by 10 men in approximately 75 minutes.

a. *Laying Out Truss Assemblies on Ground, With Wall Posts Bent Inward, and Attaching Truss Braces* (1, fig. 53). Lay out the 3 truss and wall post assemblies flat on the ground, with wall posts bent inward. Attach a truss brace to the support brackets on sides of each truss. At this point, each truss and wall post assembly will have an A-shaped appearance.

b. *Attaching Ridge Assembly to Truss Assemblies* (2, fig. 53). Extend ridge assembly, opened out, over top of truss and wall post assemblies. Raise trusses at a slight angle and attach to ridge assembly by placing spindles of truss and wall post assemblies through holes in ridge assembly.

c. *Spreading Tent Over Top, Trusses at Angle, and Attaching Guy Lines* (3, fig. 53). Spread tent over top of frame, with truss and wall post assemblies of frame at an angle. Place spindles at top of frame through grommets in tent ridge. Attach guy lines to spindles at front and rear ridge of tent.

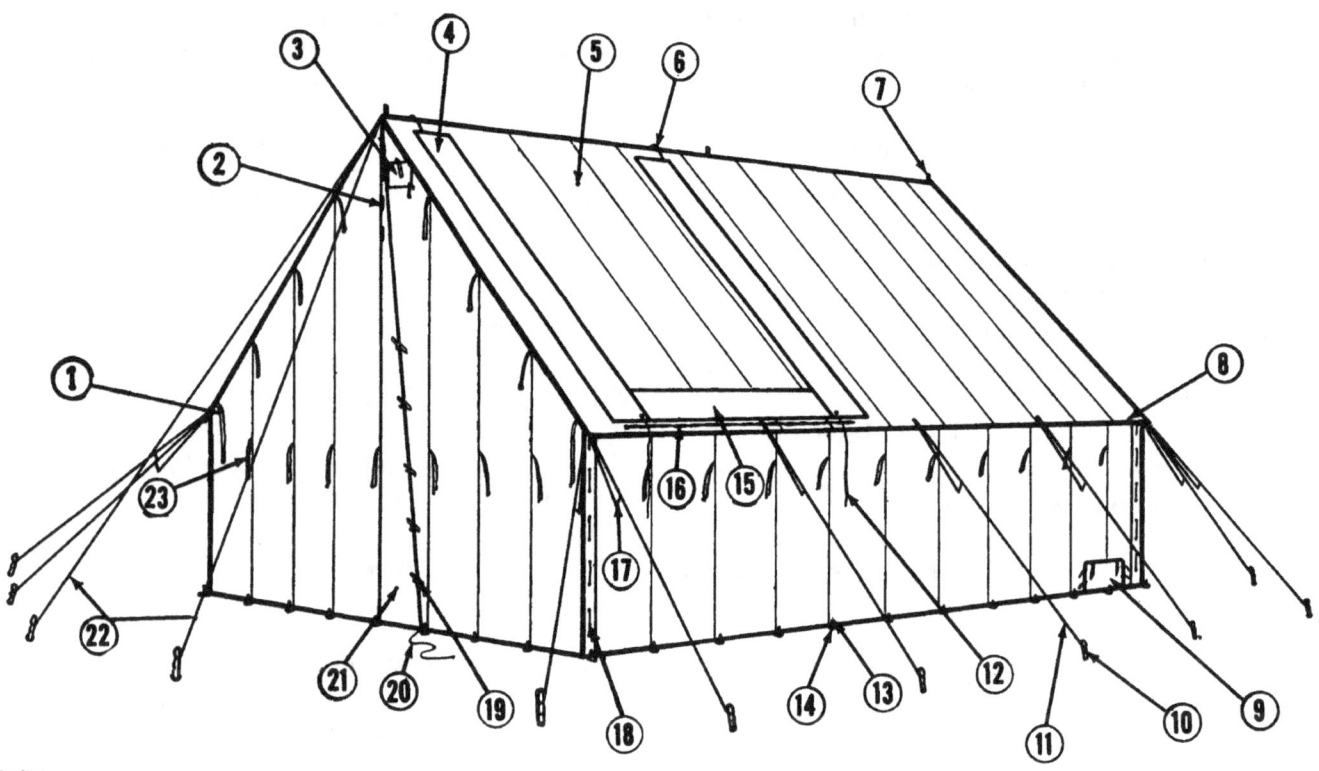

Reference No.	Part	Quantity	Federal Stock No.
1	Line, wall, 40″	4	
2	Line, tent, manila, one end sewed, ¼″ dia., 4′ 6″ long (door lacing line)	2	8340-252-2288
3	Ventilator	2	
4	Water flap	2	
5	Roof opening (cargo hatch)	1	
6	Line, tent, manila, one end sewed, ¼″ dia., 27′ long (upper draw line)	2	8340-252-2277
7	Frame, tent, maintenance shelter, armored force, complete (spindle of frame shown)	1	8340-247-4408
8	Eave		
9	Heater duct flap	2	
10	Pin, tent, wood, 24″	18	8340-261-9751
11	Line, tent, manila, one end sewed, one end w/eye, 5/16″ dia., 11′ 6″ long (eave line)	14	8340-252-2268
12	Line, tent, manila, one end sewed, ¼″ dia., 14′ long (lower draw line)	2	8340-252-2280
13	Pin, tent, wood, 16″	38	8340-261-9750
14	Line, tent, cotton, footstop, ¼″ dia., 19″ long (footstop line)	38	8340-252-2299
15	Roof opening extension flap	1	
16	Becket lacing		
17	Slip, tent line, steel	18	8340-223-8094
18	Line, tent, manila, one end sewed, ¼″ dia., 8′ long (side wall corner lacing line)	4	8340-252-2285
19	Line, tent, cotton, 3′ 4″, both ends sewed, ¼″ dia. (door fastener)	20	8340-252-2305
20	Line, tent, manila, one end sewed, ¼″ dia., 3′ 4″ long (door flap line)	4	8340-252-2291
21	Door flap	2	
22	Line, tent, manila, both ends sewed, 5/16″ dia., 64′ long (guy line)	2	8340-252-2297
23	Line, wall, 30″	48	
	Line, tent, manila, one end sewed, ¼″ dia., 4′ 6″ long (water flap line)	2	8340-252-2288
	Line, tent, manila, one end sewed, ¼″ dia., 14′ long (roof opening flap lacing line)	1	8340-252-2280
	Line, tent, manila, one end sewed, ¼″ dia., 19′ long (lacing line)	4	8340-252-2279
	Cloth, ground	6	
	Cover, tent, maintenance shelter	1	
	Line, tent, manila, one end sewed, one end w/eye, 5/16″ dia., 13′ long (cover tie line)	2	8340-252-2271

Figure 51. Tent, maintenance shelter, FWWMR, OD, complete with frame and pins, Federal Stock No. 8340-257-2557.

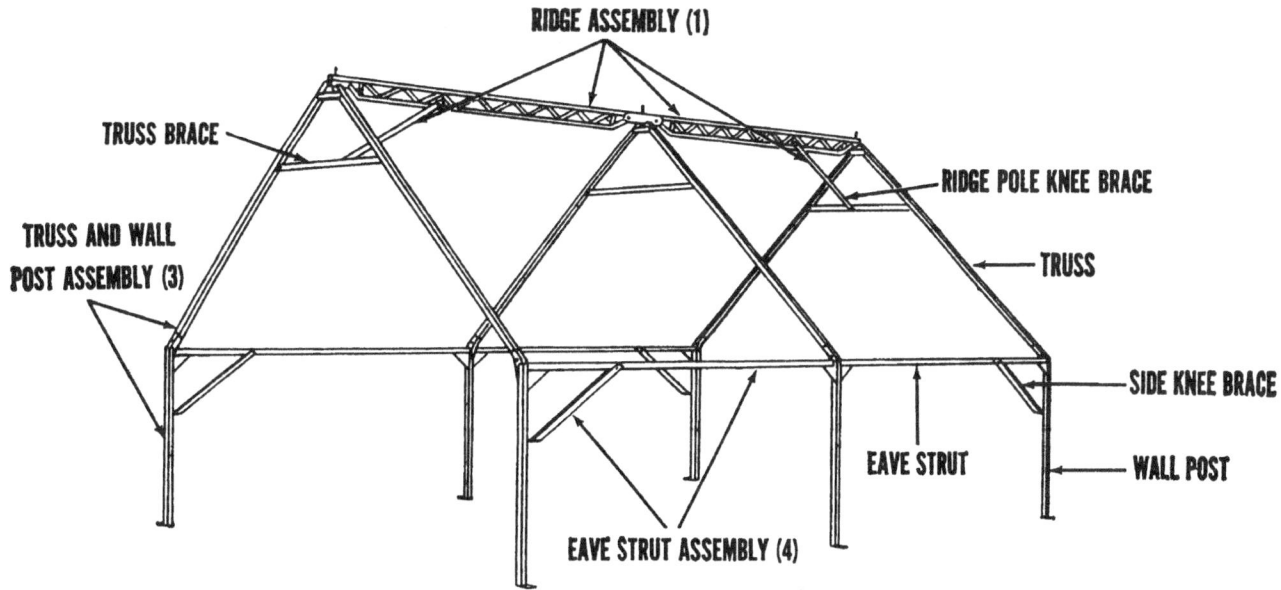

Figure 52. Frame, tent, maintenance shelter, armored force, complete, Federal Stock No. 8340-247-4408.

d. *Pushing Frame in Upright Position and Attaching Ridge Knee Braces* (4, fig. 53). Push frame into an upright position, with wall posts still bent inward. At the same time, place bolt in center of each truss brace into hole at end of each ridge knee brace.

e. *Lacing Retainers of Roof Opening Around Trusses and Ridge and Lacing Opening Flap Inside Tent* (5, fig. 53). Lace retainers of roof opening of tent around trusses and ridge of frame. Lace bottom of roof opening flap to body of tent roof.

f. *Rolling and Tying Tent Walls and Attaching and Placing Draw Lines to Slide Fasteners* (6, fig. 53).

(1) Roll up and tie side and front walls of tent.

(2) Attach lower draw lines to roof opening slide fasteners.

(3) Attach upper draw lines to roof opening slide fasteners. Place upper draw lines so that they go over tent ridge and loose ends fall on other side of tent.

(4) Close upper portion of door openings by lacing a door opening lacing line through grommets on sides of each door opening.

g. *Raising Wall Posts and Attaching Eave Struts and Side Knee Braces* (7, fig. 53).

(1) With 2 men at each of the 3 wall posts, raise one side of tent frame, making sure bolt on each truss assembly is securely inserted in slot of hinge plate on wall post. Then raise other side and lock bolts to hinge plates.

(2) Attach eave struts into position by fastening hangers at ends of eave struts around hanger brackets of truss and wall post assemblies.

(3) Fasten side knee braces of eave struts to wall posts of truss and wall post assemblies by placing bolt on angle clip of each wall post through hole at end of each side knee brace and tightening nut.

h. *Lacing Roof Opening, Securing Water Flaps, and Placing Ground Cloths* (8, fig. 53).

(1) On the inside of tent, lace bottom horizontal retainer around eave strut.

(2) On the outside of tent chain-lace beckets at bottom of roof opening extension flap through grommets on eave of tent (insert, fig. 10).

(3) On the outside of tent, close water flaps and secure them by tying water flap line at bottom of each water flap through a becket.

(4) Spread ground cloths, 3 wide and 2 deep.

i. *Rolling Down Tent Walls, Lacing Corners, and Attaching Footstops to Pins* (9, fig. 53).

(1) Untie and roll down side and front walls of tent.

(2) Close corners of tent by lacing a side wall lacing line through grommets on each side of side wall corners.

(3) Drive 16-inch pins and attach footstops.

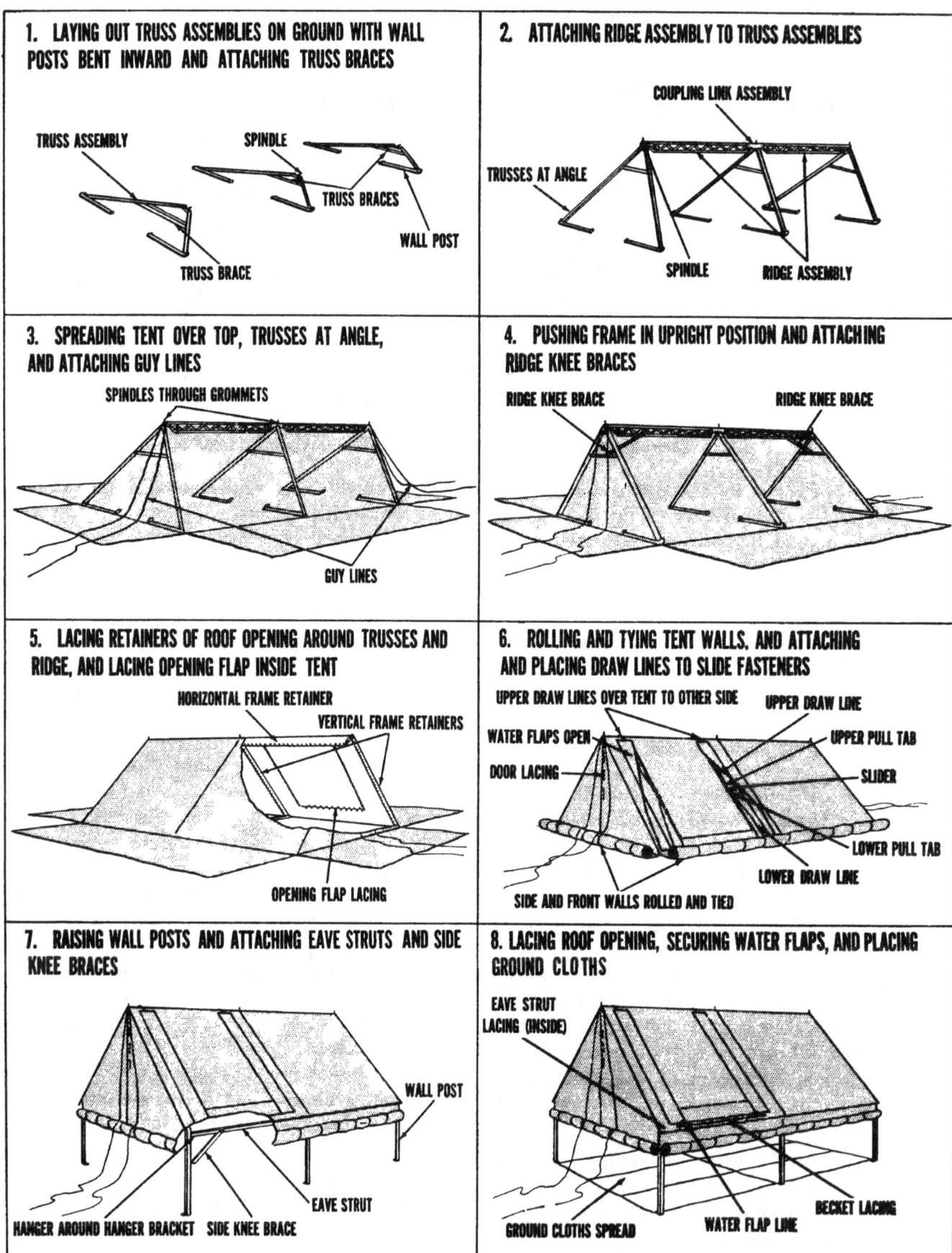

Figure 53. Steps in pitching tent, maintenance shelter, FWWMR, OD.

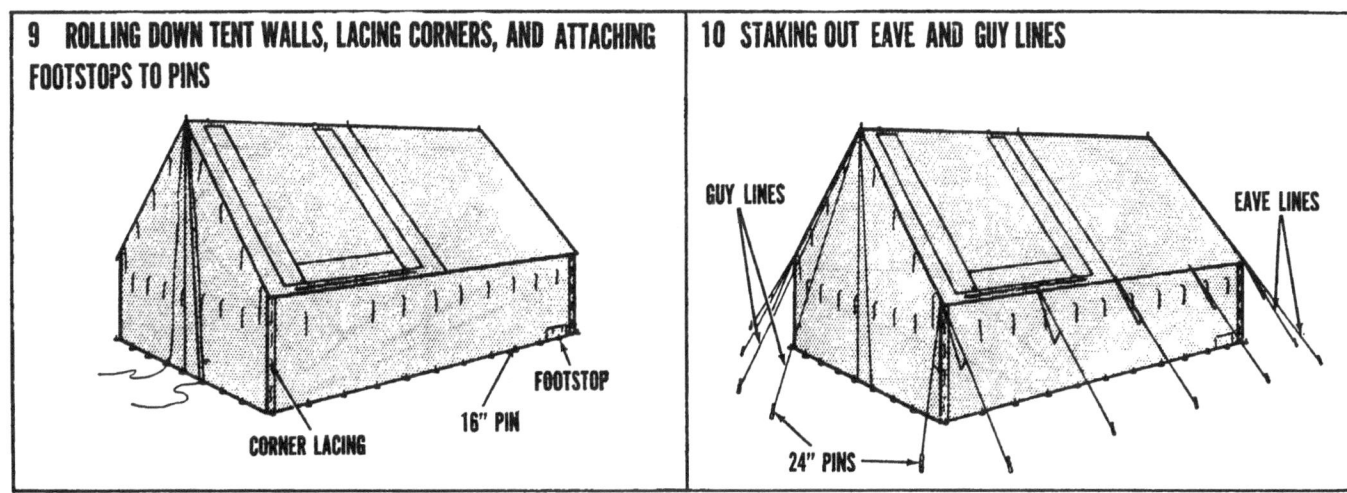

Figure 53—Continued.

j. Staking Out Eave and Guy Lines (10, fig. 53). Stake out eave and guy lines with 24-inch pins. Adjust and tighten lines.

60. Striking

a. Unfasten all eave and guy lines and remove all 24-inch pins.

b. Unfasten footstops and remove all 16-inch pins.

c. Unlace side wall corners of tent.

d. Unlace bottom horizontal retainer of tent from eave strut of frame.

e. Roll up and tie front and side walls of tent.

f. Remove eave struts and side knee braces.

g. With 1 man at each of the 3 wall posts, lower one side of tent frame by bending wall posts inward; then lower other side.

h. Untie front and side walls of tent.

i. Untie lace lines of tent and remove all lashings from around the various sections of frame structure.

j. Open front and rear doors of tent and slide canvas off ridge of tent frame. Carry canvas completely away from frame.

k. Separate frame by reversing procedure used to erect it.

61. Folding and Packing

a. Folding Tent.

(1) *Folding at ridge* (1, fig. 54). Fold tent at ridge, with sides and two parts of front and rear ends together. Roll guy lines up and tie. Coil eave lines toward center.

(2) *Placing ground cloths* (2, fig. 54). Place ground cloths on tent just above eave, staggering them to use space to the best advantage. Turn sod cloth in toward eave.

(3) *Folding front and rear ends over* (3, fig. 54). Fold front and rear ends over toward center.

(4) *Folding ridge at deck to eave* (4, fig. 54). Fold ridge at deck to eave.

(5) *Folding deck over side walls, then in half* (5, fig. 54). Fold deck over side walls, and then in half. Coil eave lines toward center.

(6) *Folding ends toward center, end over end, and placing in cover* (6, fig. 54). Fold ends toward center, end over end, and place in cover. Close cover, folding long flap first, then short flap. Tie cover with 2 tie lines.

b. Packing Frame (fig. 55). Pack sections of frame in 3 crates as follows:

(1) *Crate 1.*
  1 ridge assembly
  2 eave strut assemblies
  3 truss braces

(2) *Crate 2.*
  1 truss and wall post assembly
  2 eave strut assemblies

(3) *Crate 3.*
  2 truss and wall post assemblies

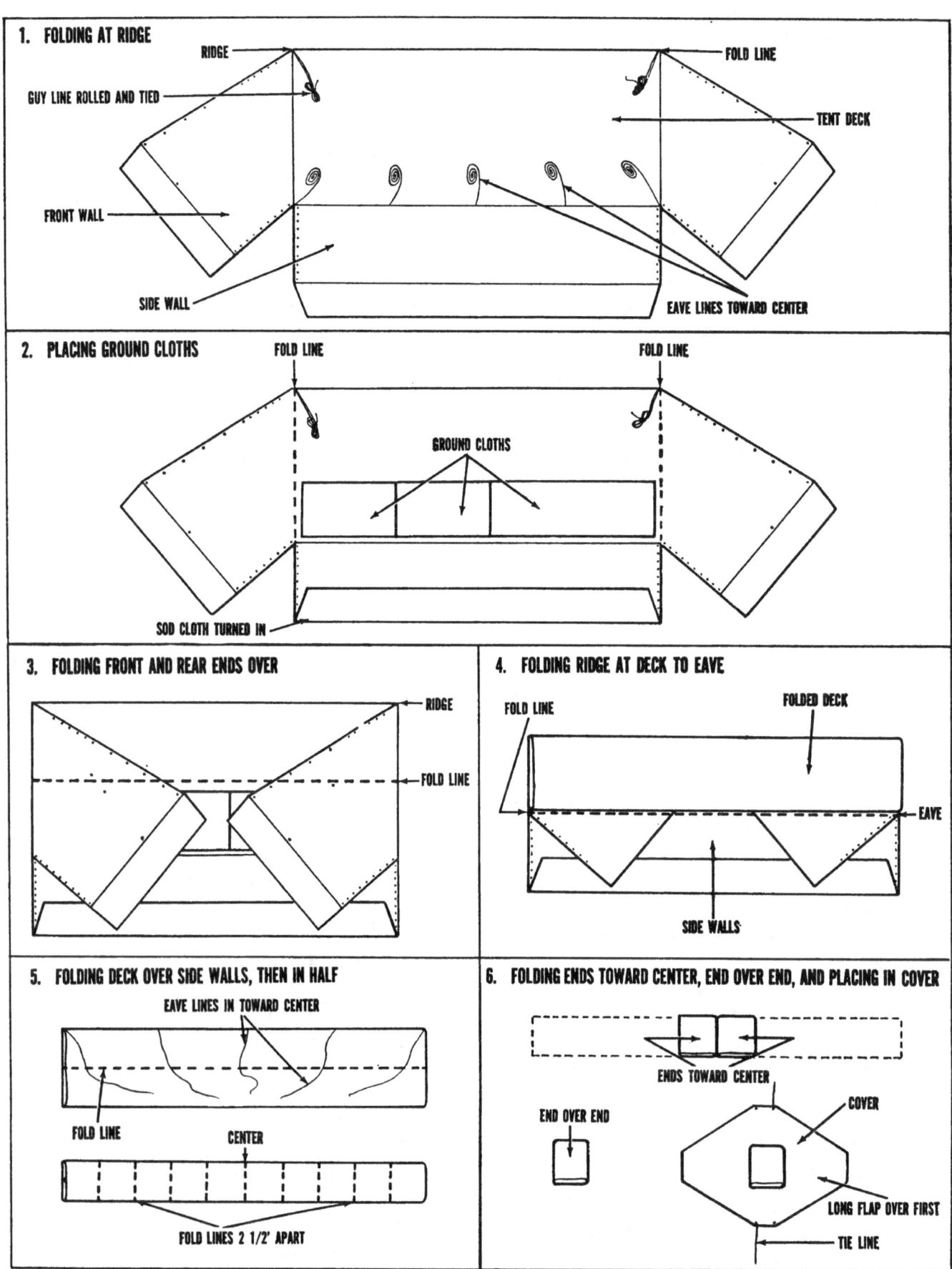

Figure 54. Steps in folding tent, maintenance shelter, FWWMR, OD.

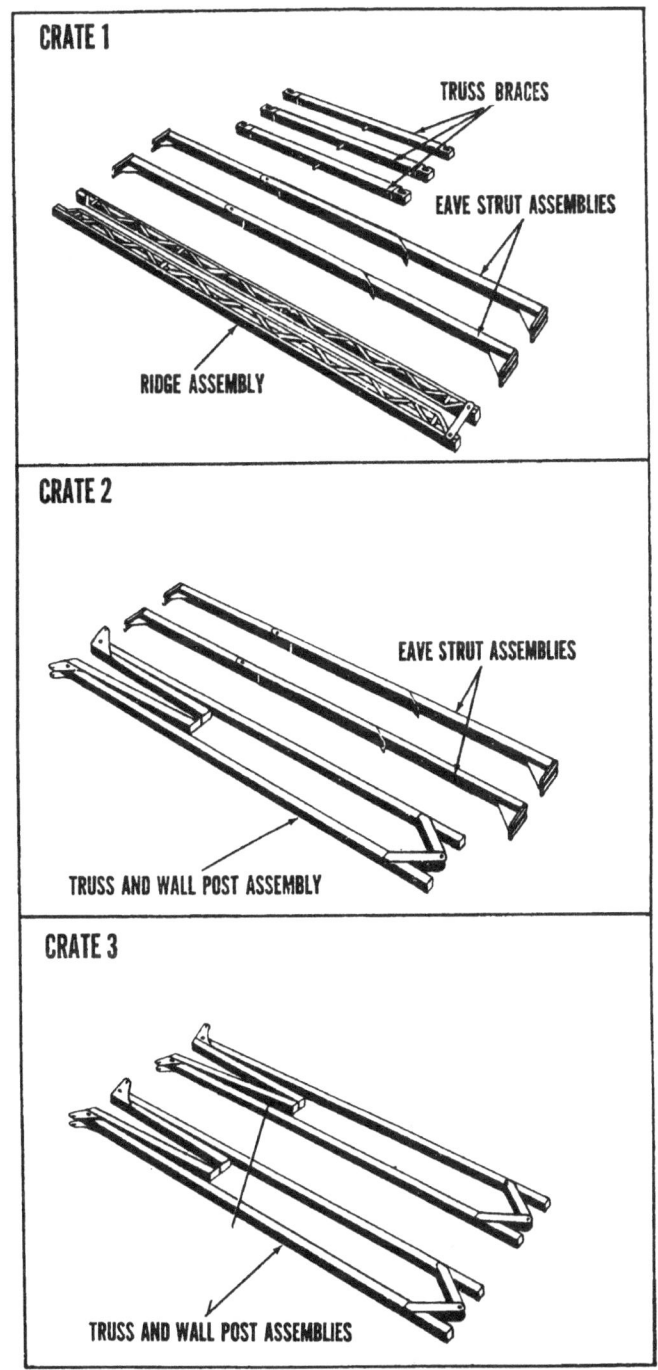

Crate 1. Frame section, tent, maintenance shelter, section No. 1 bottom, Federal Stock No. 8340-242-7871
Crate 2. Frame section, tent, maintenance shelter, section No. 2 middle, Federal Stock No. 8340-242-7870
Crate 3. Frame section, tent, maintenance shelter, section No. 3 top, Federal Stock No. 8340-242-7869

Figure 55. Packing frame, tent, maintenance shelter, Armored Force, complete, Federal Stock No. 8340-247-4408.

Section XI. MOUNTAIN TENT

62. Use

The tent, mountain, 2-man, FWWMR, OD and white, complete with pins and poles (fig. 56), is a lightweight tent for two men, used in cold-climate operations, particularly in mountainous areas, where ordinary means of transportation are not available for bringing in heavier types of tentage.

63. Description

The tent is triangular in cross section, with an entrance at each end.

a. Dimensions. The tent is 54 inches wide, 82 inches long, and 43 inches high. A wall height of 12 inches, formed by stretching eave lines on each side, gives the tent a pitch of 31 inches.

b. Weight and Cubage. The tent weighs 6 pounds and the pins and poles weigh 3⅝ pounds. The tent has a cubage in storage of 0.5 cubic feet, and the pins and poles have a cubage of 0.2 cubic feet.

c. Floorspace. The floorspace is approximately 30.75 square feet.

d. Materials. The tent sides are made of lightweight cotton cloth. The tent floor is made of synthetic fiber, vinyl resin or synthetic rubber coated, cloth. The tent is white and olive drab, reversible; it may be camouflaged by exposing the appropriate color.

e. Entrances.

(1) The tent has 2 tubular tunnel entrances, 27 inches in diameter and 24 inches long.

(2) A tunnel entrance may be closed by tying it either from the inside or outside with tie tapes. To tie entrances, wind tie tape around tunnel entrance as if entrance were the mouth of a bag, and fasten it with a half hitch.

(3) A tunnel entrance may be kept open by pulling it out and securing it to a guy line with tie tapes, or it may be rolled against the tent and secured by tying the

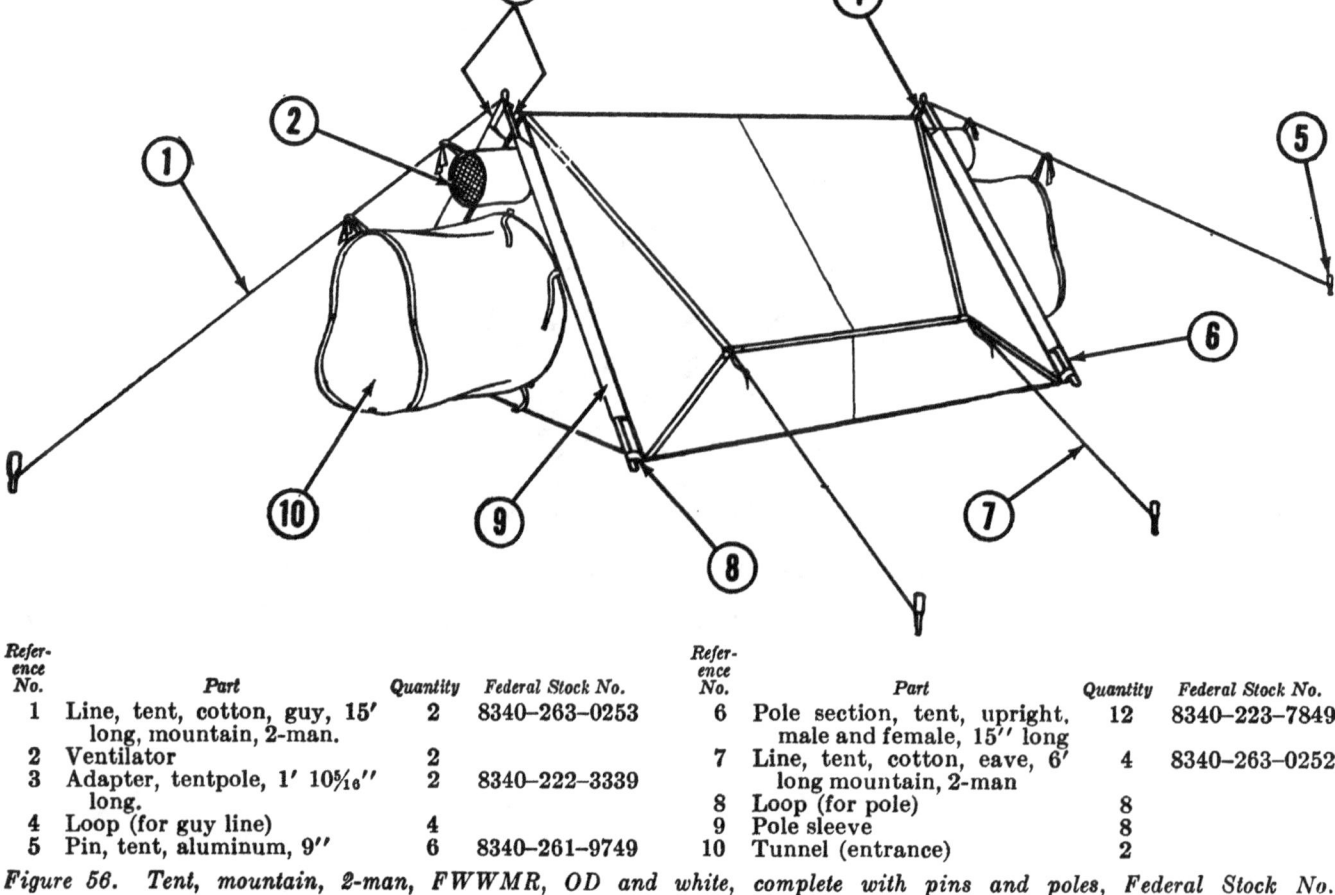

Reference No.	Part	Quantity	Federal Stock No.
1	Line, tent, cotton, guy, 15' long, mountain, 2-man.	2	8340-263-0253
2	Ventilator	2	
3	Adapter, tentpole, 1' 10 5/16" long.	2	8340-222-3339
4	Loop (for guy line)	4	
5	Pin, tent, aluminum, 9"	6	8340-261-9749
6	Pole section, tent, upright, male and female, 15" long	12	8340-223-7849
7	Line, tent, cotton, eave, 6' long mountain, 2-man	4	8340-263-0252
8	Loop (for pole)	8	
9	Pole sleeve	8	
10	Tunnel (entrance)	2	

Figure 56. Tent, mountain, 2-man, FWWMR, OD and white, complete with pins and poles, Federal Stock No. 8340-254-9017.

tie tapes on the tent through grommets on the outside opening of the entrance.
- (4) Tubular mosquito netting, attached to the body of the tent inside the entrance tunnels, may be closed by tying it tight either from the inside or outside by tie tapes. To tie the mosquito netting, wind tie tapes around opening of netting as if it were the mouth of a bag, and fasten it with a half hitch.

f. Ventilation.
- (1) Ventilation is of the greatest importance in the mountain tent, because the cloth has been coated to make it impermeable. The tent may be ventilated by opening the tunnel entrances or by using the built-in ventilators.
- (2) An 8-inch-in-diameter ventilator, with mosquito netting at the outside opening, is at each end of the tent. In good weather, the ventilators are kept wide open by tying them to the guy lines with tie tapes. In storms, they are left hanging loosely to provide adequate protection as well as ventilation. The ventilators should never be closed when a gasoline-burning stove is lighted. In cold weather, there is an additional reason for leaving the ventilators open. Unless the moisture caused by breathing and cooking can pass off into the outside air, it forms as frost on the roof of the tent. In a wind, this shakes off and wets the clothes and sleeping bags.

g. Floor. The floor is constructed as an integral part of the tent. Special care should be taken not to tear the floor with boots.

64. Pitching

Two men can pitch the tent in approximately 10 minutes.

a. Preliminary Procedures (1, fig. 57).
- (1) Spread tent on ground in position it is to occupy, with desired color on outside: olive drab in normal situations and white under snowy conditions. To reverse tent for proper color, pull inside of tent through one of the entrance tunnels, taking care not to damage the fabric.
- (2) Assemble tentpoles so that 4 poles of 3 sections each are made. Place poles on ground alongside the 2 tentpole adapters.

b. Inserting Poles Through Loops and Sleeves and Attaching Adapters (2, fig. 57).
- (1) Insert tentpoles through corner loops and pole sleeves of tent.
- (2) Attach pole adapters to tentpoles.

c. Raising Front End of Tent (3, fig. 57).
- (1) Raise front tentpoles and adapter to a position so that front end of tent is vertical.
- (2) Place front guy line through ring of adapter and stake guy line out to a pin in front of tent.

d. Raising Rear End of Tent (4, fig. 57).
- (1) Raise rear tentpole and adapter to a position so that ridge of tent is almost level and rear end of tent is vertical.
- (2) Place rear guy line through ring of adapter and stake guy line out to a pin to rear of tent.

e. Tying Ventilator and Entrance Tie Tapes to Guy Lines and Staking Out Eave Lines (5, fig. 57).
- (1) Tie ventilator and tunnel entrance tie tapes to guy lines.
- (2) Attach eave lines to the 2 loops on each side of tent and stake eave lines out to pins.

f. Anchoring Corner. When additional anchorage of the tent is required, lines may be attached to the corner loops and secured to the guy-line pins.

g. Pitching Tent Without Poles and Pins. In order to achieve maximum mobility, the tent may be pitched without using poles and pins. This procedure is especially valuable in wooded terrain. The corners of the tent and the front and rear guy lines may be staked down with available sticks or stones. If the ridge of the tent sags, it may be supported by attaching a line to the loop in the center of the ridge and securing the line to a tree. Skis and ski poles may be used in place of tentpoles and pins. Although the tent may be pitched without pins and poles, these items should always be available.

h. Pitching Tent in Rocky Terrain. In rocky terrain, it may be impossible to drive the tent pins into the ground. In this case, attach guy lines to rocks.

i. Pitching Tent in Loose and Powdery Snow. When the snow on which the tent is pitched is loose and powdery, the guy lines may be attached to ski poles or ice axes, which are driven down into the snow after it has been packed; or the lines may

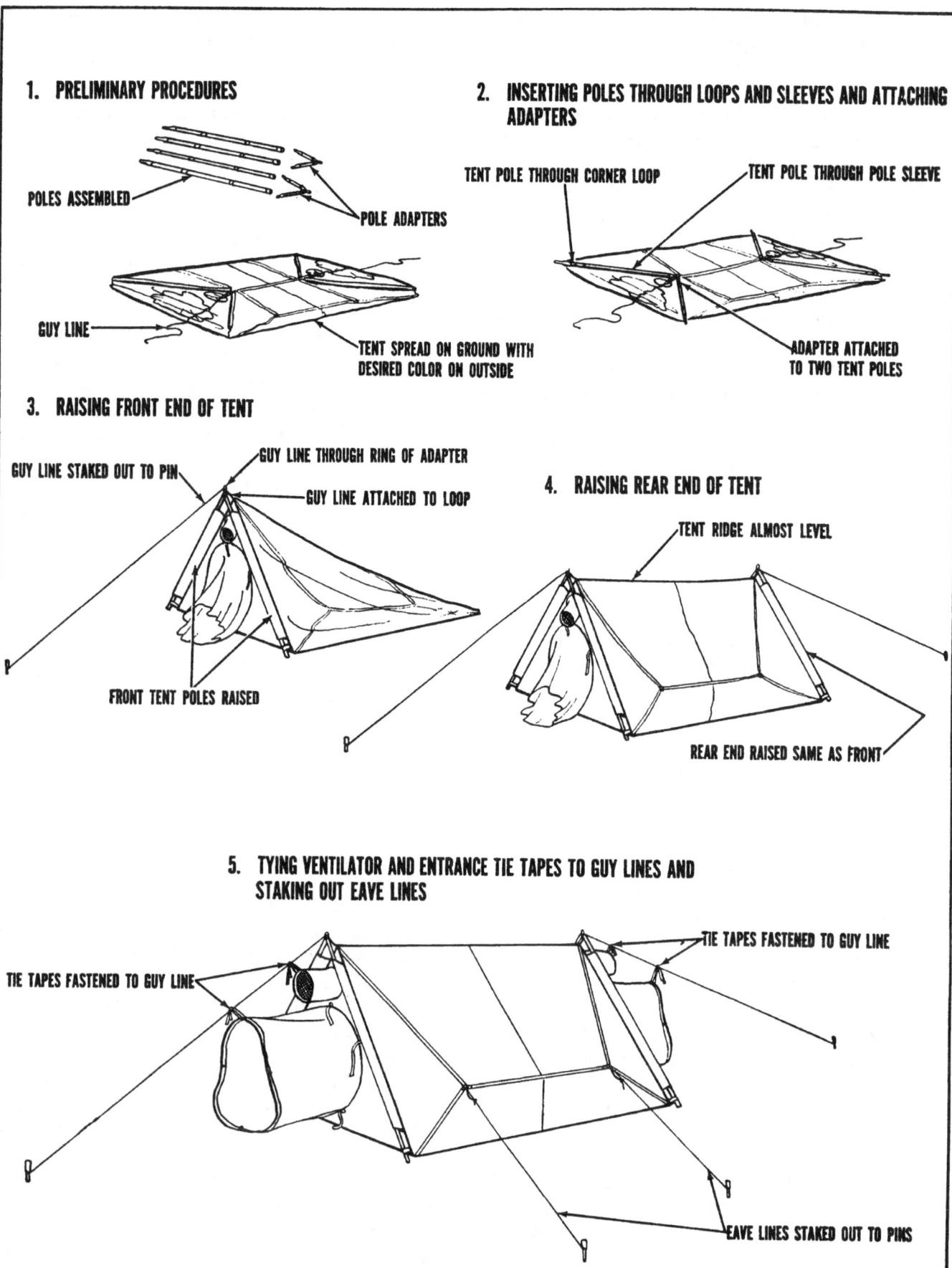

Figure 57. Steps in pitching tent, mountain, 2-man, FWWMR, OD and white.

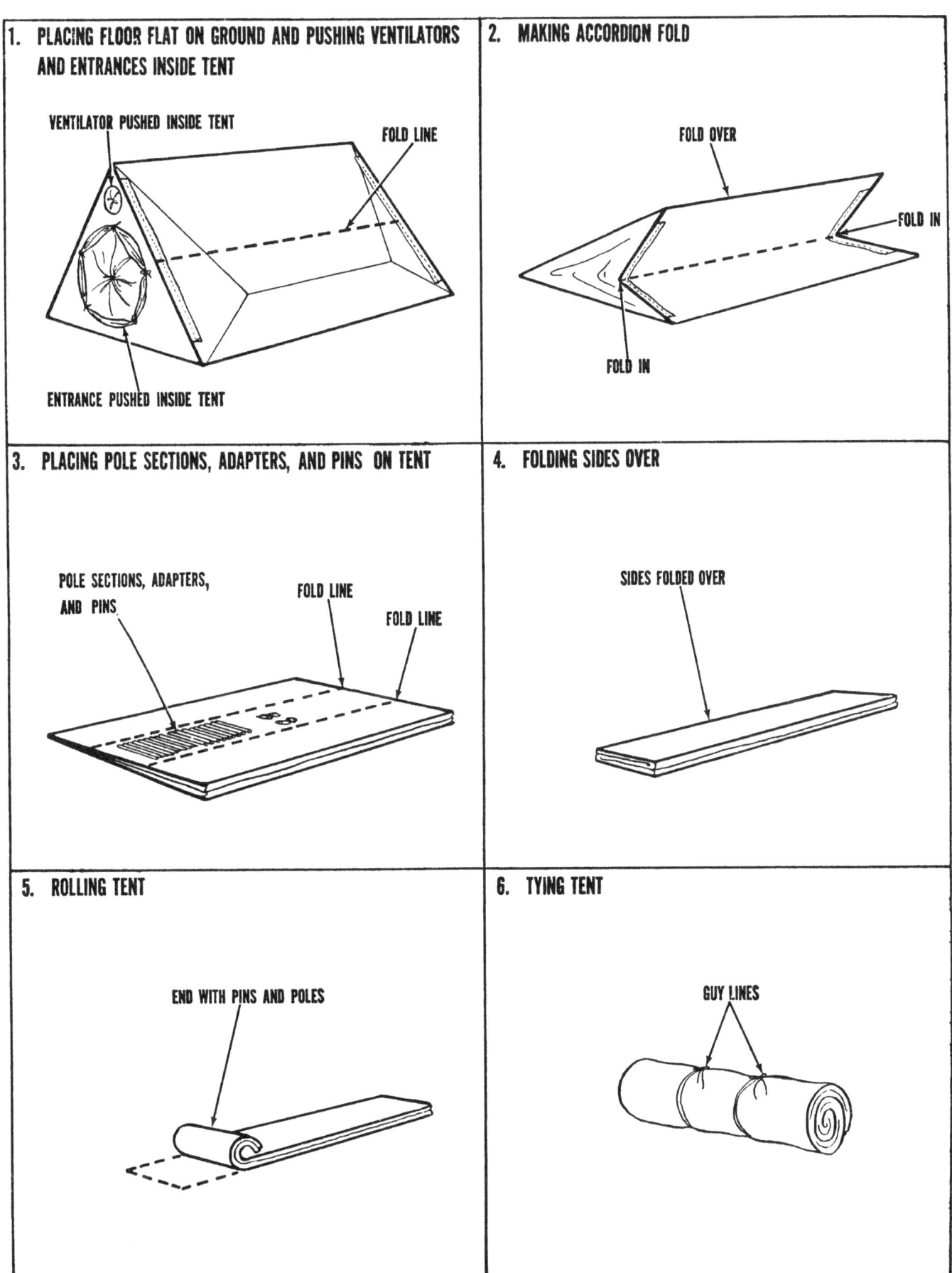

Figure 58. Steps in folding and rolling tent, mountain, 2-man, FWWMR, OD and white.

be attached to a "dead man" anchor. This is made by burying a tent pin or stick horizontally in a hole in the snow and stamping the snow on top of the anchor until it is thoroughly packed.

65. Striking

a. Untie ventilator and tunnel entrance tie tapes from guy lines.

b. Remove guy and eave lines from pins.

c. Remove pins from ground.

d. Untie guy lines from webbing loops at front and rear peaks of tent.

e. Unfasten adapters from poles and remove poles from tent; then disassemble poles.

66. Folding and Rolling

a. Placing Floor Flat on Ground and Pushing Ventilators and Entrances Inside Tent (1, fig. 58). Place tent so that bottom is flat on ground. Push ventilators and tunnel entrances inside tent.

b. Making Accordion Fold (2, fig. 58). With one man at each end of the tent, make an accordion fold by folding one side of tent inward at center and folding the other side over so that it covers bottom of tent.

c. Placing Pole Sections, Adapters, and Pins on Tent (3, fig. 58). Place pole sections, adapters, and pins at center of one end of folded tent. Eave line should remain attached.

d. Folding Sides Over (4, fig. 58). Fold sides of folded tent over toward center.

e. Rolling Tent (5, fig. 58). Starting at the end with pins and poles, roll folded tent tightly toward the other end.

f. Tying Tent (6, fig. 58). Tie rolled tent with 2 guy lines. The tent may now be placed on the pack or stored.

Section XII. RED CROSS MARKERS

67. Large Red Cross Marker

a. Use. The large Red Cross marker (panel marker set, Red Cross, cotton duck, 100 feet long, 100 feet wide, Federal Stock No. 8345-00-19030) is used by the Medical Corps. It is designed to be spread flat on the ground to indicate a hospital area.

b. Description.

(1) The marker is composed of five 20-foot-wide sections made of 9.68-ounce vinyl-coated cotton duck.

(2) The marker weighs 1,225 pounds.

(3) The marker has the following component parts:

Part	Quantity	Federal Stock No.
Panel marker, Red Cross, cotton duck, end section, 100 x 20 ft.	2	8345-00-19029
Panel marker, Red Cross, cotton duck, intermediate section, 100 x 20 ft.	2	8345-00-19027
Panel marker, Red Cross, cotton duck, center section, 100 x 20 ft.	1	8345-00-19028
Pin, marker, Red Cross, canvas, 100 x 100 ft., 12 in. long	275	8345-00-19015
Roll, pin, marker, Red Cross, canvas, 100 x 100 ft.	5	8345-00-19016
Cover, section, with pins, marker, Red Cross, canvas, 100 x 100 ft.	5	8345-00-19005

c. Pitching. Spread sections on ground in proper sequence: end section, intermediate section, center section, intermediate section, and end section. Fasten sections together by inserting pins through overlapping grommets of sections and through grommets around edge of marker so that a large red cross on a white field is formed.

d. Striking. Remove pins and separate sections.

e. Folding.

(1) Fold each of the 5 sections separately. Fold each section twice toward center along its long dimension. Then, in 2½-foot folds, fold ends of each section toward center. Place each folded section on a cover.

(2) Place the 55 pins provided with each section in a pin roll and close roll securely, tying with tie tape. Place each pin roll, with pins, on top of a folded section.

(3) Close flaps of each cover securely and tie tie lines tightly through grommets, making sure that no part of marker is exposed.

68. Small Red Cross Marker

a. Use. The small Red Cross marker (panel marker set, Red Cross, cotton duck, 9 ft. 6 in. long, 21 ft. 6 in. wide, Federal Stock No. 8345-00-19031, fig. 59) is used by the Medical Corps. It is designed to be lashed down over the ridge of an A-shaped tent (usually a sectional hospital tent or a general purpose tent) to indicate the use of that tent for medical purposes.

b. Description.

(1) The marker is made of 9.68-ounce vinyl-coated cotton duck, bearing two red crosses on a white field.

(2) The marker weighs 25 pounds and has a cubage in storage of 1.5 cubic feet.

(3) The marker is provided with center guy lines, corner guy lines, and tent slips so that it can be lashed down securely:

Part	Quantity	Federal Stock No.
Line, tent, manila, both ends sewed, 5/16" dia., 50' long (center guy line)	2	8340-252-2293
Line, tent, manila, one end sewed, one end w/eye, 3/8" dia., 15' long (corner guy line)	8	8340-252-2270
Slip, tent line, steel	12	8340-223-8094

c. *Pitching.*

(1) Before pitching tent, spread Red Cross marker over roof of tent on which it is to be displayed. Place center of marker over ridge of tent so that one of the Red Cross insignia is on each side of tent roof and one end of each guy line falls to each side of tent.

(2) Pitch tent according to instructions.

(3) Spread each side of marker out smooth by adjusting corner lines.

(4) Attach corner guy lines and center guy lines to pins staked out for tent lines. Tighten and adjust corner guy lines and center guy lines by adjusting tent slips.

d. *Folding.* Remove the 2 center guy lines and the 8 corner guy lines from pins. Spread marker flat on ground and coil all lines except 2 of the corner guy lines at one end of marker. Fold marker twice toward center along its long dimension. Then, in 2½-foot folds, fold ends toward center, and tie with the 2 corner guy lines.

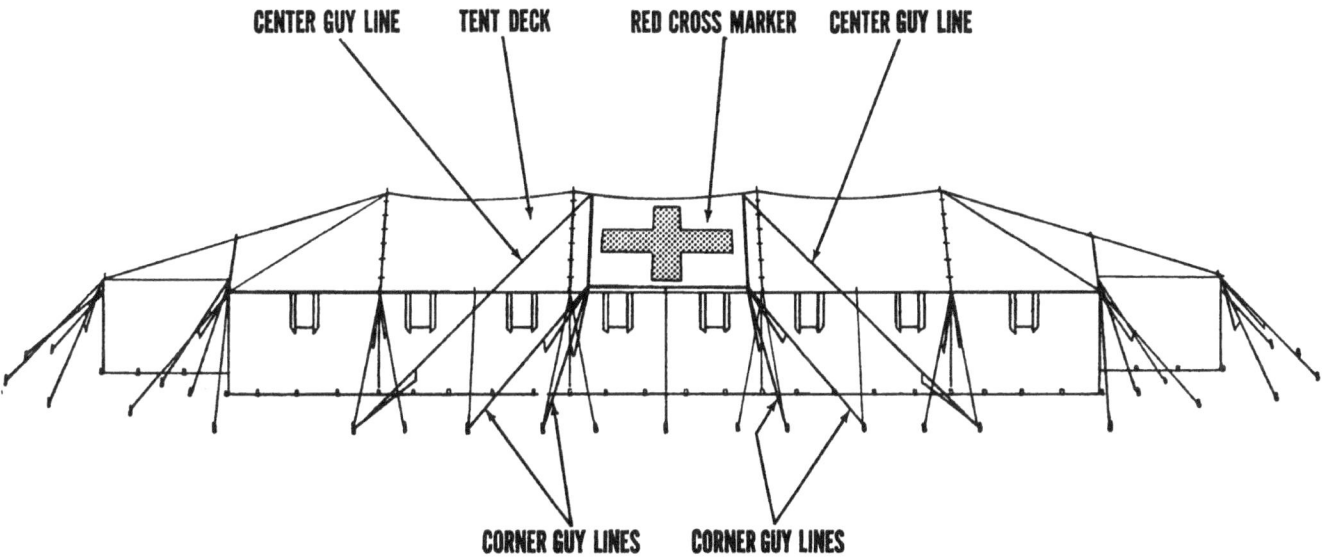

Figure 59. Small Red Cross marker (panel marker set, Red Cross, cotton duck, 9 ft. 6 in. long, 21 ft. 6 in. wide), attached to tent, FWWMR, hospital sectional.

Section XIII. WALL TENT, LARGE

69. Use

The tent, wall, large, FWWMR, OD, complete with pins and poles (fig. 60), is used primarily to provide office shelter for staff sections, usually at divisional level; it can accommodate 12 men and the necessary folding tables and office equipment. It may also be used for the storage of supplies, for an officers' mess serving about 20 men, or for quartering personnel. It can shelter 6 men when a stove is installed and 8 men when a stove is not installed.

70. Description

The tent is an A-shaped square-end rectangular tent which comes in one section.

a. *Dimensions.* The tent is 14 feet 6 inches wide and 14 feet long. The ridge height is 11 feet and the wall height is 4 feet 6 inches, giving a pitch of 6 feet 6 inches.

b. *Weight and Cubage.* The tent weighs 130 pounds, and the pins and poles weigh 145 pounds. The tent has a cubage in storage of 5.8 cubic feet, and the pins and poles have a cubage of 3.1 cubic feet.

c. *Floorspace.* The floorspace is approximately 203 square feet.

d. *Materials.* The 12.29-ounce cotton duck is used on all parts of the tent except the sod cloth, which is 9.85-ounce cotton duck.

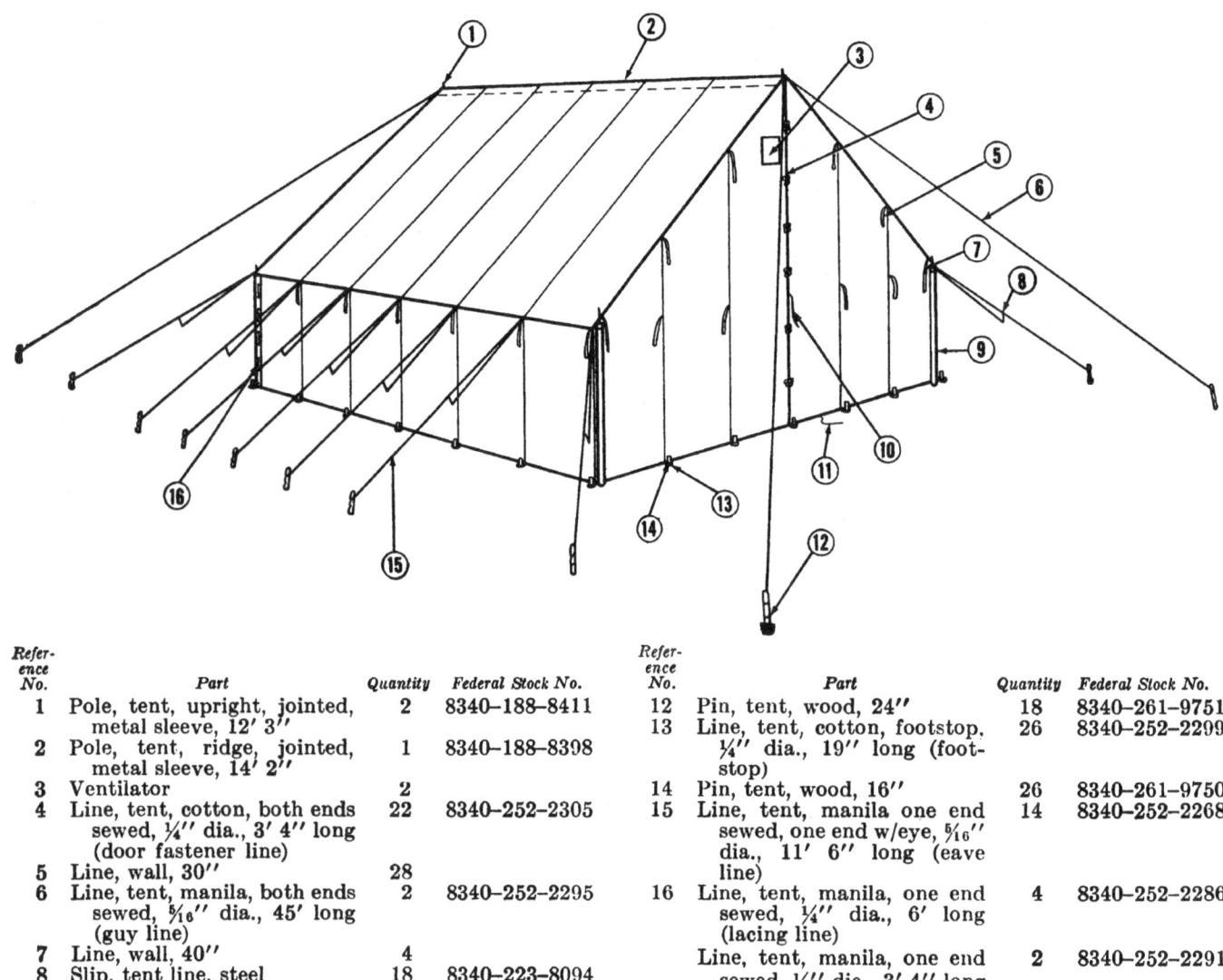

Reference No.	Part	Quantity	Federal Stock No.
1	Pole, tent, upright, jointed, metal sleeve, 12' 3"	2	8340-188-8411
2	Pole, tent, ridge, jointed, metal sleeve, 14' 2"	1	8340-188-8398
3	Ventilator	2	
4	Line, tent, cotton, both ends sewed, ¼" dia., 3' 4" long (door fastener line)	22	8340-252-2305
5	Line, wall, 30"	28	
6	Line, tent, manila, both ends sewed, 5⁄16" dia., 45' long (guy line)	2	8340-252-2295
7	Line, wall, 40"	4	
8	Slip, tent line, steel	18	8340-223-8094
9	Pole, tent, upright, solid, 4' 9"	4	8340-188-8403
10	Line, tent, cotton, both ends sewed, ¼" dia., 3' 4" long (door wall line)	2	8340-252-2305
11	Line, tent, manila, one end sewed, ¼" dia., 3' 4" long (door flap line)	2	8340-252-2291
12	Pin, tent, wood, 24"	18	8340-261-9751
13	Line, tent, cotton, footstop, ¼" dia., 19" long (footstop)	26	8340-252-2299
14	Pin, tent, wood, 16"	26	8340-261-9750
15	Line, tent, manila one end sewed, one end w/eye, 5⁄16" dia., 11' 6" long (eave line)	14	8340-252-2268
16	Line, tent, manila, one end sewed, ¼" dia., 6' long (lacing line)	4	8340-252-2286
	Line, tent, manila, one end sewed, ¼" dia., 3' 4" long (extension cloth line)	2	8340-252-2291
	Cover, tent, wall, large	1	
	Line, tent, manila, one end sewed, one end w/eye, 5⁄16" dia., 13' long (cover tie line)	2	8340-252-2271

Figure 60. Tent, wall, large, FWWMR, OD, complete with pins and poles, Federal Stock No. 8340-257-2548.

e. Extension Cloth. There is an extension cloth attached to one end of the tent. It is used for joining two tents together.

f. Doors. Doors are formed by slits in the middle of each end wall, the halves of which are overlapped. Doors may be tied with door fastener lines.

g. Ventilation. Ventilator openings are provided at the front and rear walls of the tent near the ridge. The openings are protected by canvas flaps.

h. Heating. The tent is heated by an M-1941 tent stove. Either of the ventilator openings may be used as a stovepipe opening.

i. Cover. The tent is provided with a cover for use when in storage or when being transported.

j. Fly. A fly (Federal Stock No. 8340-188-9030) is available as a separate item of issue. The fly may be suspended above the deck of the tent to lower the temperature within the tent. It may also be pitched independently of the tent to provide quick shade and shelter, for example as a field kitchen. The fly is often erected against the rear of a kitchen truck. The cooking is done in the truck and the food served under the fly. The fly measures 21 feet 6 inches by 14 feet 5 inches. It weighs 50 pounds and has a cubage in storage of 1.6 cubic feet. Six tent lines (line, tent, manila, one end sewed, one end with eye, 5/16 in. dia., 11 ft. 6 in. long, Federal Stock No. 8340-252-2268) accompany the fly; they are used as eave lines.

71. Ground Plan

Before pitching large wall tent, study ground plan carefully (fig. 61).

72. Pitching

a. Pitching Tent. Four men can pitch the tent in approximately 30 minutes.

   (1) *Spreading tent, closing corners and doors, driving pins, and attaching corner footstops to pins* (1, fig. 62).

      (*a*) Spread tent on ground with eave corners matching bottom corners.

      (*b*) Close corners of tent by lacing the four 6-foot lacing lines through grommets on ends of side and end wall corners, and tie at both bottom and eave. Close doors by tying door fastener lines through grommets.

      (*c*) Draw corners of tent out and line them up square.

      (*d*) Drive 16-inch pins at corners and attach footstops to pins.

   (2) *Placing eave poles in position, assembling ridge pole and upright center poles,*

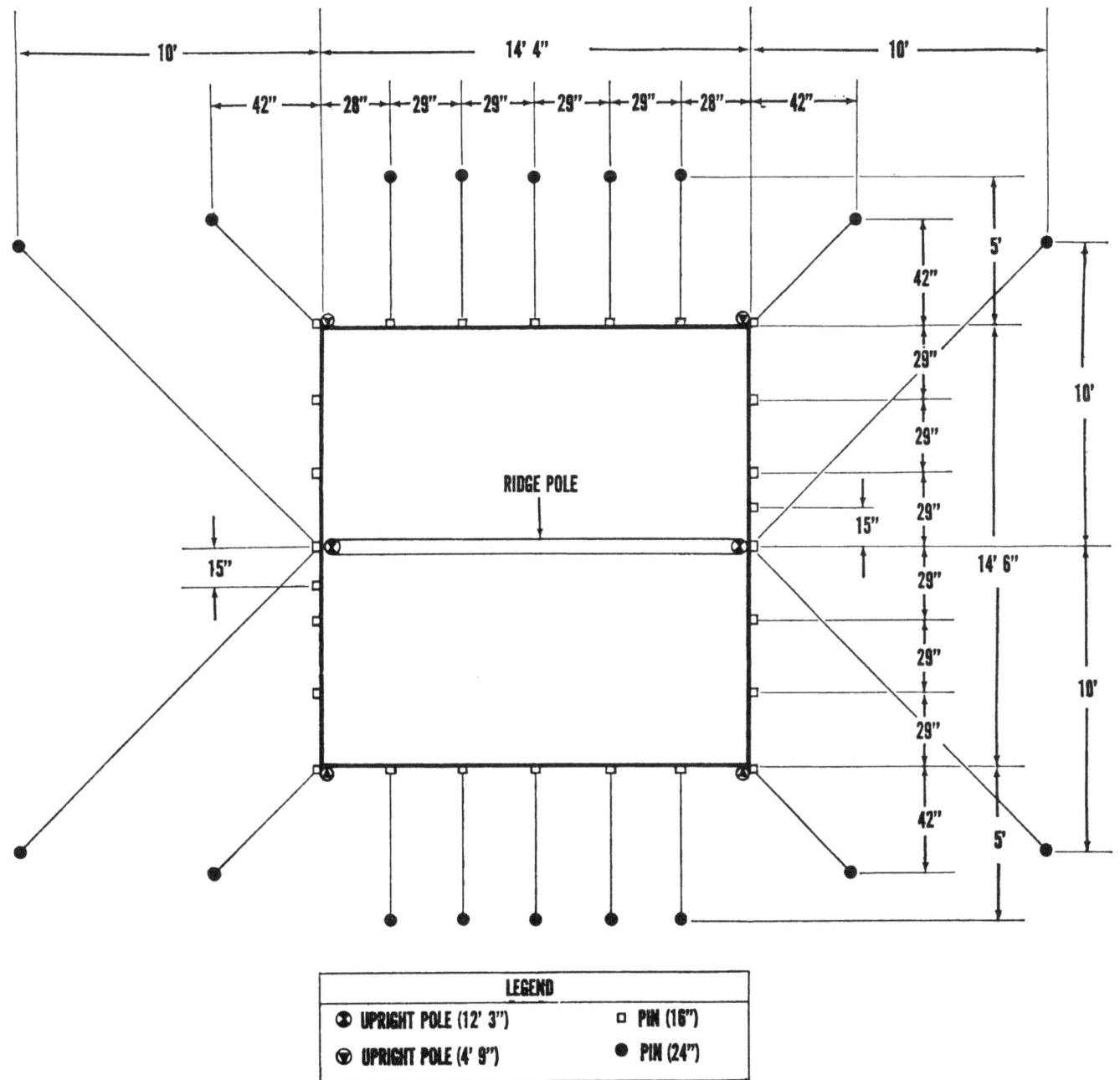

Figure 61. Ground plan of tent, wall, large, FWWMR, OD.

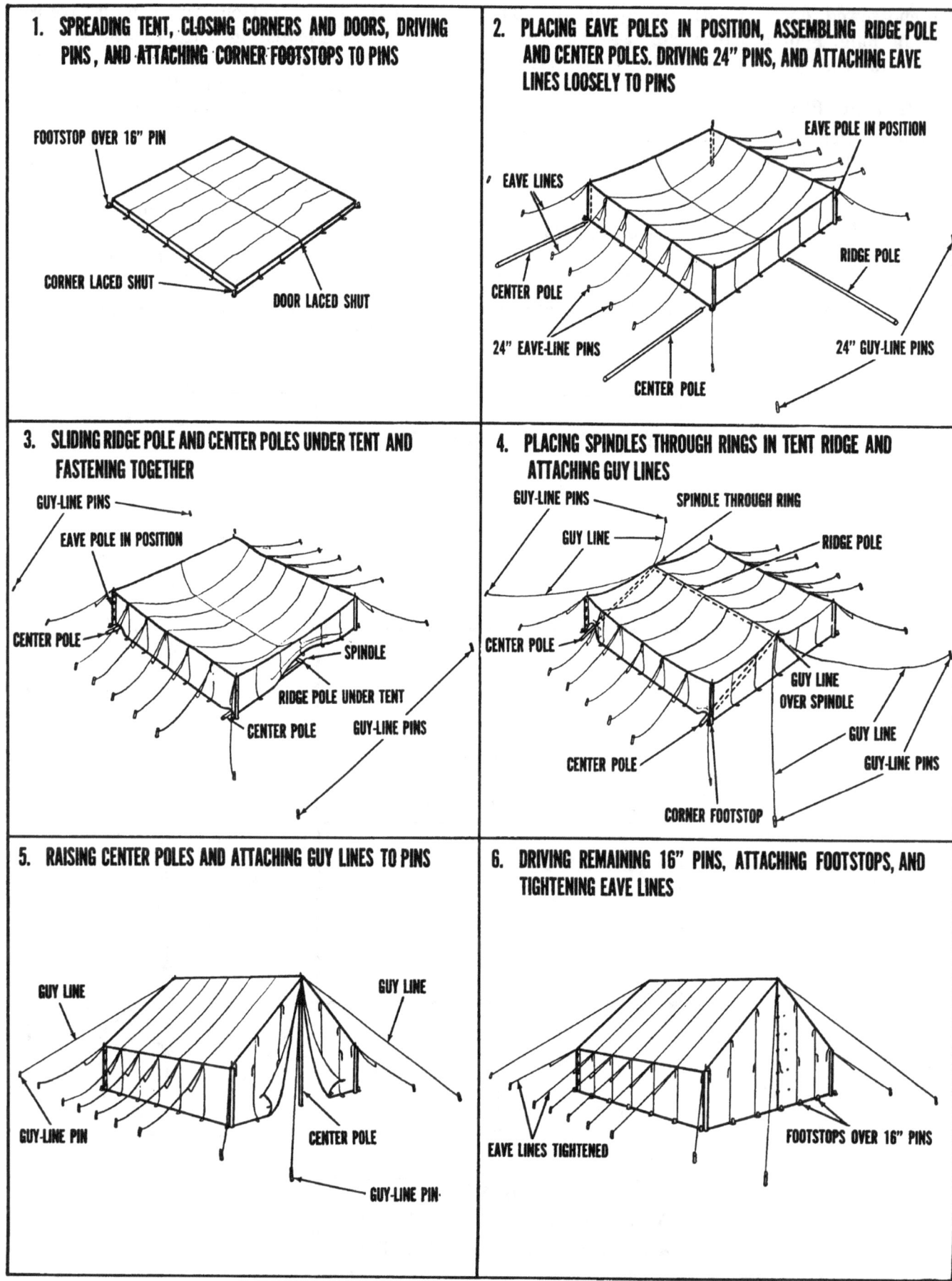

Figure 62. Steps in pitching tent, wall, large, FWWMR, OD.

driving 24-inch pins, and attaching eave lines loosely to pins (2, fig. 62).

(a) Place the 4 corner 4-foot 9-inch upright eave poles in position and insert spindle of each pole through the ¾-inch hand-worked ring in tent canvas at eave.

(b) Assemble the 14-foot 2-inch ridge pole and the 12-foot 3-inch upright center poles and place them in position ready to slide under tent wall.

(c) Drive the 24-inch eave-line pins 5 feet out from tent side. Drive the 24-inch guy-line pins according to ground plan.

(d) Attach eave lines loosely to the 24-inch eave-line pins.

(3) *Sliding ridge pole and center poles under tent and fastening together* (3, fig. 62). Slide ridge pole and upright center poles under tent and fasten together by inserting spindles of upright poles through holes in ridge pole, with round side of ridge pole up.

(4) *Placing spindles through rings in tent ridge and attaching guy lines* (4, fig. 62).

(a) Remove corner footstops from the 16-inch pins.

(b) Raise center upright poles slightly and place spindles of poles through rings in canvas of tent ridge.

(c) Attach guy lines to tent by placing cut splice in center of each guy line over spindle of each upright pole outside tent.

(5) *Raising center poles and attaching guy lines to pins* (5, fig. 62).

(a) Dig a 12-inch hole for each of the 2 center poles in positions indicated by the ground plan. With 2 men at each center pole, raise poles to a vertical position and place them in holes. Then tamp earth around each pole.

(b) Attach guy lines to pins and tighten.

(6) *Driving remaining 16-inch pins, attaching footstops, and tightening eave lines* (6, fig. 62).

(a) Drive the remaining 16-inch pins and attach footstops to all 16-inch pins.

(b) Tighten all eave lines.

b. *Joining Two Tents Together.* When 2 tents are to be joined together, erect the 2 tents end to end about 5 inches apart, with extension cloth at end of one tent adjacent to end without extension cloth of the second tent. Fasten extension cloth at end of one tent to end of the other tent by placing hand-worked rings at ridge and at eaves of extension cloth over spindles of upright poles at end of the other tent. Then on each side near the base, tie an extension cloth line through grommet on extension cloth of one tent and through grommet on canvas of the other tent.

c. *Pitching Fly With Tent* (fig. 63).

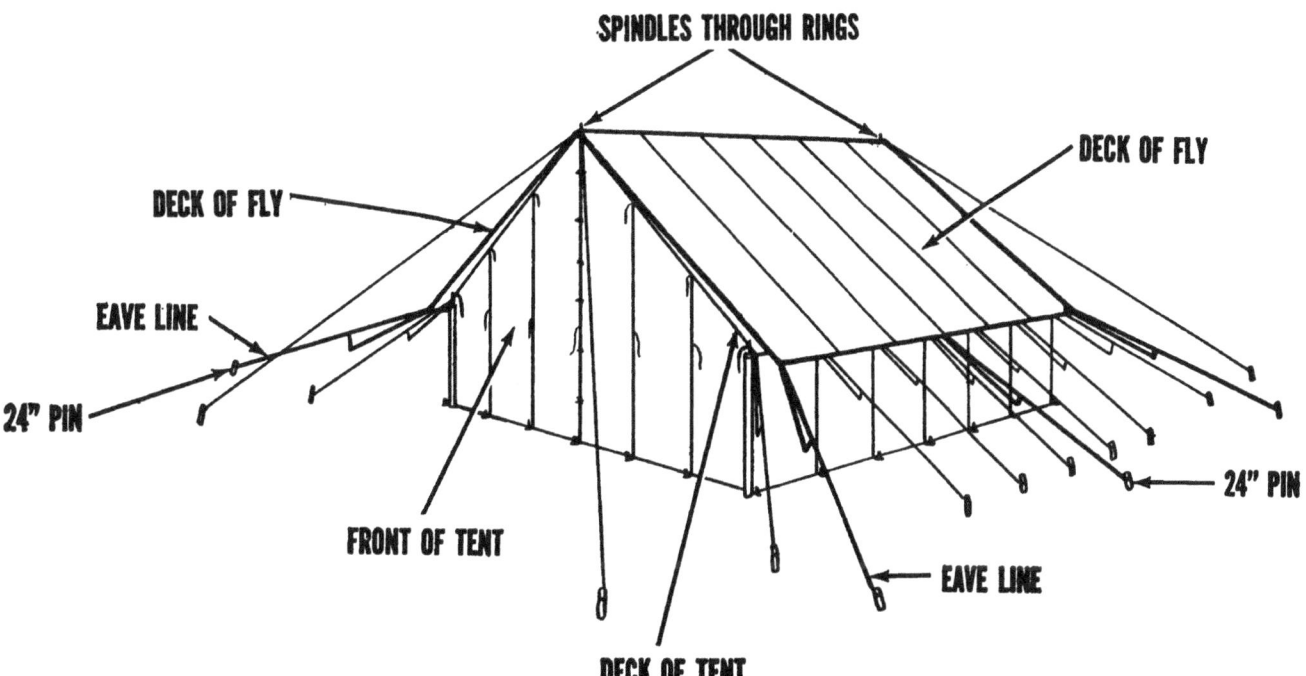

Figure 63. Pitching fly with tent, wall, large, FWWMR, OD.

(1) The fly, tent, FWWMR, wall, large, OD, may be pitched together with the tent.

(2) To pitch the fly with the tent, spread fly over tent, which has been spread out on the ground, so that deck of fly covers deck of tent and rings in ridge of fly match rings in ridge of tent.

(3) Erect tent and fly together by following instructions for pitching tent (*a* above). Before raising tent, be sure to place rings in ridge of fly over spindles of upright center poles of tent above tent ridge.

(4) After tent has been raised into position and tent eave and guy lines have been attached to pins and tightened, attach the 6 eave lines of fly to 24-inch pins staked out 8 feet from tent. First, attach to pins the 4 corner eave lines diagonally out from tent; then the 2 side eave lines straight out from tent. The 24-inch pins are available as separate items of issue; they do not come as component parts of the tent or of the fly.

(5) Six 4-foot 9-inch upright poles may be used to provide a greater airspace between the fly and the tent. When poles are used, place spindle of each pole through inner eye of double-eye bar at

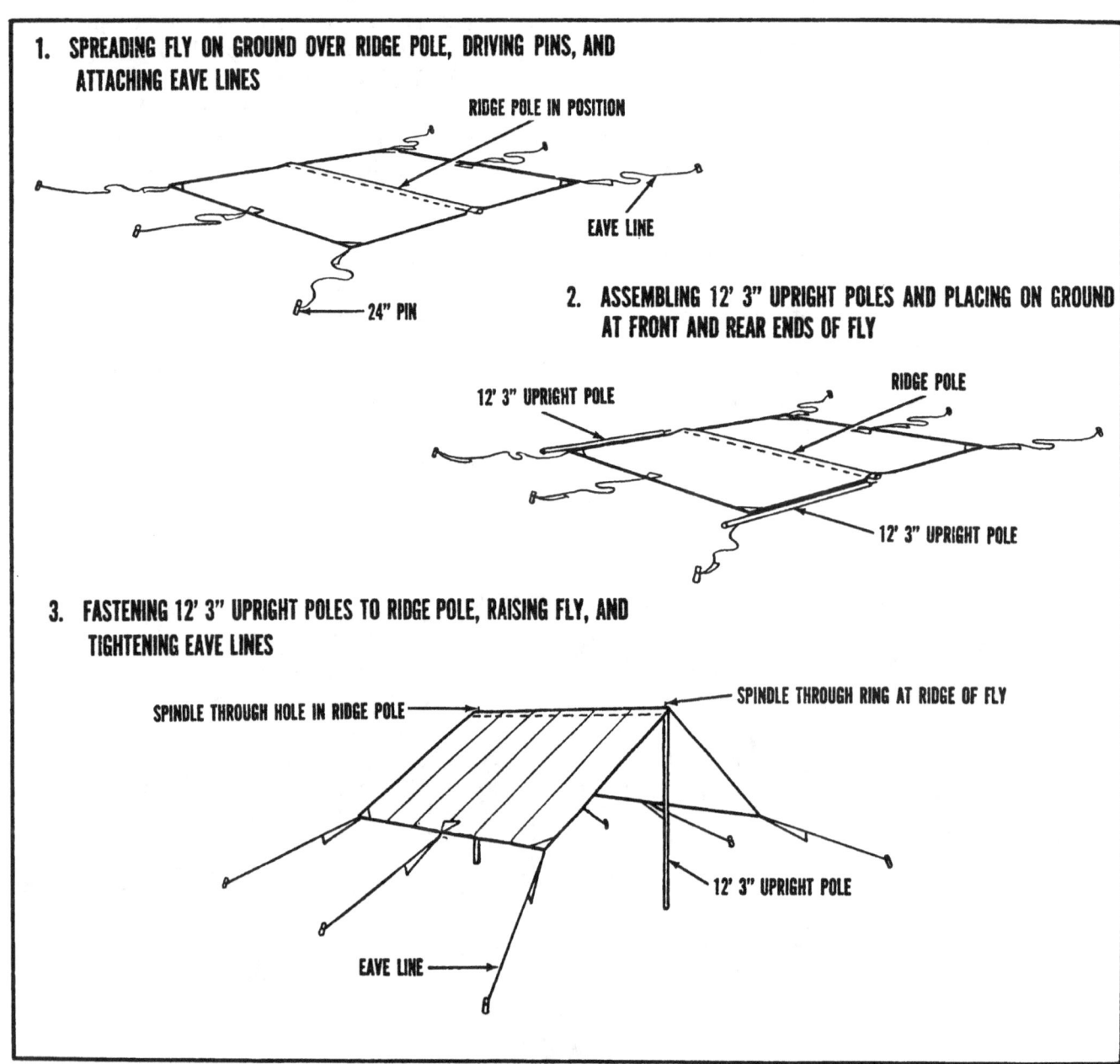

Figure 64. Steps in pitching fly, tent, FWWMR, wall, large, independently of tent.

corners and sides of fly at eave. The 4-foot 9-inch poles are available as separate items of issue; they do not come as component parts of the tent or of the fly.

d. Pitching Fly Without Tent. The fly, FWWMR, tent, wall, large, OD, can be pitched independently of the tent. The 14-foot 2-inch ridge pole, the two 12-foot 3-inch upright poles, and the six 24-inch pins used in pitching the fly independently of the tent are available as separate items of issue; they do not come as component parts of the fly.

(1) *Spreading fly on ground over ridge pole, driving pins, and attaching eave lines* (1, fig. 64).

(*a*) Place ridge pole on ground and spread fly out so that ridge of fly is directly over ridge pole.

(*b*) Drive the 24-inch pins 5 feet out from fly, the 4 corner pins diagonally out and the 2 side pins straight out.

(*c*) Attach eave lines loosely to the 24-inch pins.

(2) *Assembling 12-foot 3-inch upright poles and placing on ground at front and rear ends of fly* (2, fig. 64). Assemble the two 12-foot 3-inch upright poles and place them on the ground, one at the front and one at the rear end of fly, with spindle of each pole toward middle of each end of fly.

(3) *Fastening 12-foot 3-inch upright poles to ridge pole, raising fly, and tightening eave lines* (3, fig. 64).

(*a*) Fasten the two 12-foot 3-inch upright poles to ridge pole by placing spindle of each upright pole through hole at each end of ridge pole, with round side of ridge pole up.

(*b*) Place spindle of each 12-foot 3-inch upright pole through ring in canvas at ridge of fly.

(*c*) With 2 men at each upright pole, raise poles to a vertical position.

(*d*) Adjust and tighten all eave lines.

(4) *Pitching fly without ridge pole.* The fly may be pitched without a ridge pole. In this case, two 45-foot guy lines (manila, both ends sewed, $5/16$-inch dia) and four additional 24-inch pins will be required. These are available as separate items of issue. Before raising fly, place guy lines in position by placing center of each guy line over spindle of each 12-foot 3-inch upright pole outside fly.

(5) *Pitching fly with eave poles.* The fly may be pitched with eave poles. In this case, six 4-foot 9-inch upright poles, available as separate items of issue, will be required. Before driving the 24-inch eave-line pins, place spindle of each 4-foot 9-inch upright pole through inner eye of double-eye bar at corners and sides of fly at eave.

73. Striking

Four men can strike the tent in 20 minutes.

a. Unfasten the 4 corner lacing lines.

b. Remove all footstops from 16-inch pins.

c. Remove all lines, except the 4 corner eave lines, from the 24-inch pins.

d. Lower the two 12-foot 3-inch upright center poles, lowering the 14-foot 2-inch ridge pole and removing from tent.

e. Remove the 4 corner poles from under tent, and lower tent.

f. Remove the 4 corner eave lines from the 24-inch pins.

g. Remove all pins from ground.

h. Disassemble the two 12-foot 3-inch upright center poles and the ridge pole.

74. Folding

a. Folding at Ridge (1, fig. 65). Spread tent flat on ground and fold extension cloth over body of tent. Fold tent at ridge so that bottoms of side walls are even and the two parts of front and rear ends are together. Roll the 45-foot guy lines up and tie. Coil eave lines toward center.

b. Folding Sod Cloths Over (2, fig. 65). Fold sod cloths over body of tent.

c. Folding Front and Rear Ends Over (3, fig. 65). Fold front and rear ends over toward center.

d. Folding Ridge at Deck to Eave (4, fig. 65). Fold ridge at deck to eave.

e. Folding Deck Over Side Walls, Then in Half (5, fig. 65). Fold deck over side walls and then in half. Make sure eave lines are coiled in toward center.

f. Folding Ends Toward Center, End Over End, and Placing in Cover (6, fig. 65). Fold ends toward center, end over end, and place in cover. Close cover, folding long flap first, then short flap. Tie cover with two tie lines.

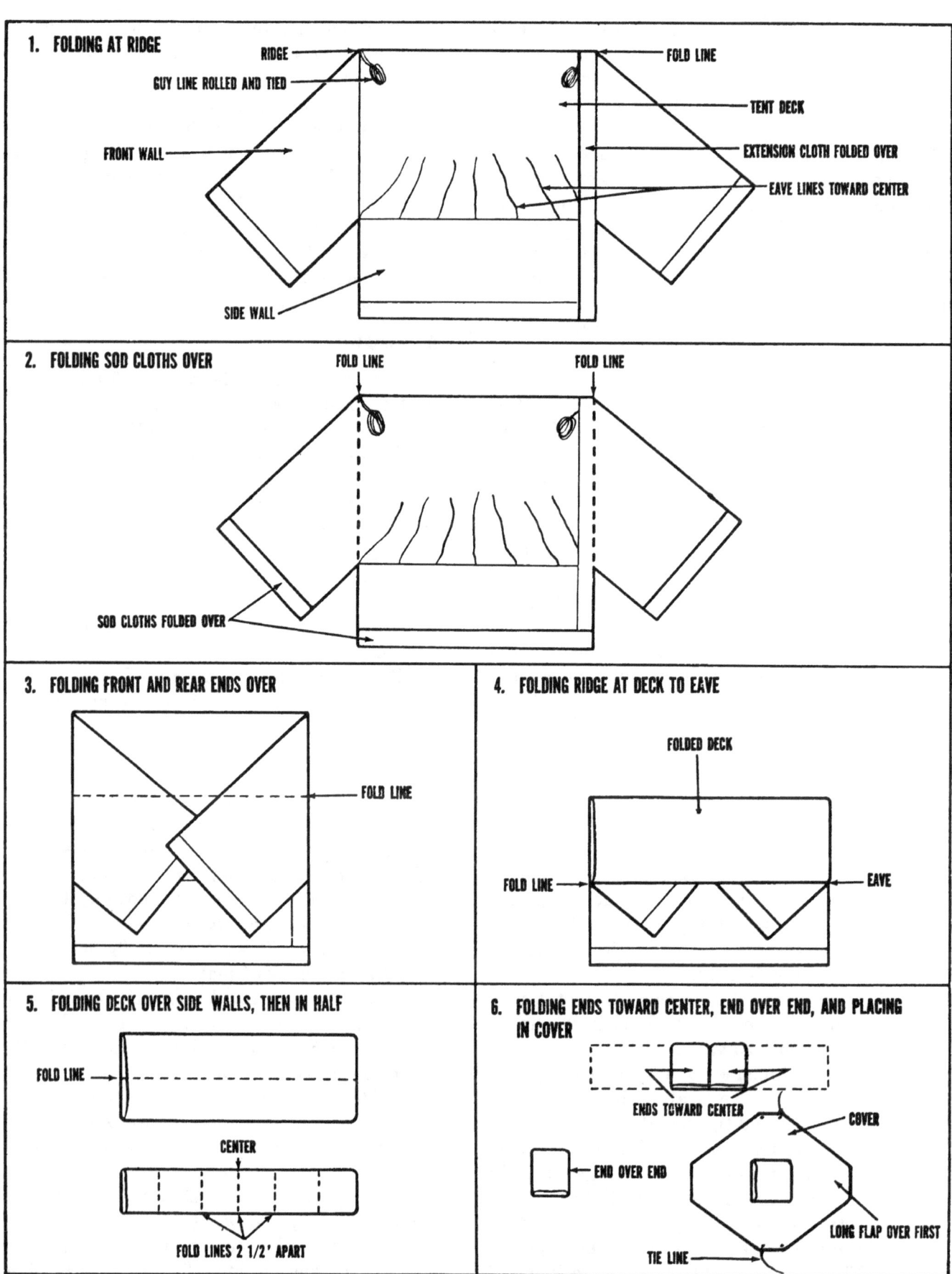

Figure 65. Steps in folding tent, wall, large, FWWMR, OD.

Section XIV. WALL TENT, SMALL

75. Use

The tent, wall, small, FWWMR, OD, complete with pins and poles (fig. 66), is used mainly for the shelter of officers when in the field and not in combat. It will accommodate 2 men. It may also be used as a field first-aid station, command post, or small storage tent.

76. Description

The tent is an A-shaped square-end rectangular tent.

a. Dimensions. The tent is 8 feet 10 inches wide and 9 feet 2 inches long. The ridge height is 8 feet 6 inches and the wall height is 3 feet 9 inches, giving a pitch of 4 feet 9 inches.

b. Weight and Cubage. The tent weighs 55 pounds, and the pins and poles weight 60 pounds. The tent has a cubage in storage of 3.4 cubic feet, and the pins and poles have a cubage of 4.1 cubic feet.

c. Floorspace. The floorspace is approximately 81 square feet.

d. Materials. The top, side walls, and all reinforcements are made of 12.29-ounce duck, and the sod cloth is made of 9.85-ounce duck. The tent comes in one section.

e. Doors. Doors are formed by slits in the middle of each end section, the halves of which are overlapped. Doors may be tied with door fastener lines.

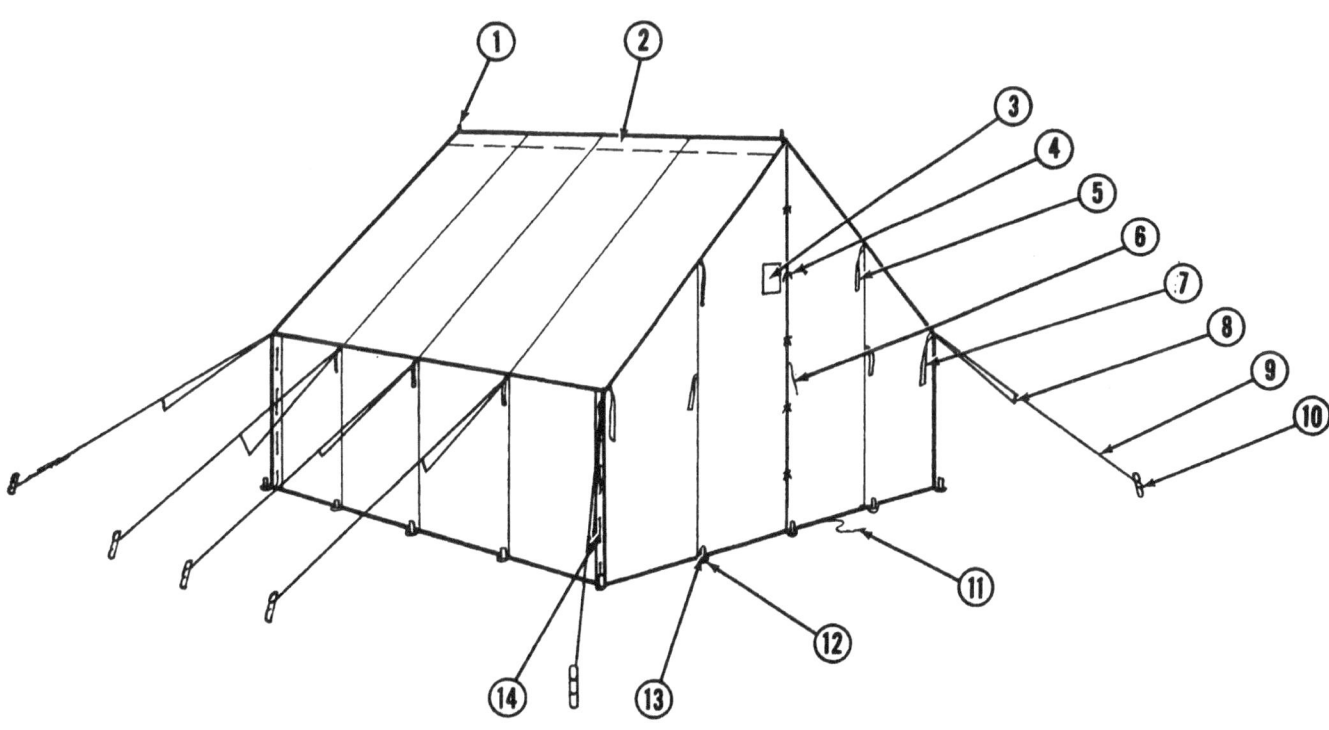

Reference No.	Part	Quantity	Federal Stock No.
1	Pole, tent, upright, solid, 9'	2	8340-188-8410
2	Pole, tent, ridge, solid, 9'	1	8340-188-8394
3	Ventilator	2	
4	Line, tent, cotton, both ends sewed, ¼" dia., 3' 4" long (door fastener)	18	8340-252-2305
5	Line, wall, 30"	16	
6	Line, tent, cotton, both ends sewed, ¼" dia., 3' 4" long (door wall line)	2	8340-252-2305
7	Line, wall, 40"	4	
8	Slip, tent line, steel	10	8340-223-8094
9	Line, tent, manila, one end sewed, one end w/eye, ¼" dia., 8' long (eave line)	10	8340-252-2269
10	Pin, tent, wood, 24"	10	8340-261-9751
11	Line, tent, manila, one end sewed, ¼" dia., 3' 4" long (door flap line)	2	8340-252-2291
12	Line, tent, cotton, footstop, ¼" dia., 19" long (footstop)	18	8340-252-2299
13	Pin, tent, wood, 16"	18	8340-261-9750
14	Line, tent, manila, one end sewed, ¼" dia., 4' 6" long (lacing line)	4	8340-252-2288
	Cover, tent, FWWMR, wall, small	1	8340-242-4069
	Line, tent, manila, one end sewed, one end w/eye, ¼" dia., 8' long (cover tie line)	2	8340-252-2269

Figure 66. Tent, wall, small, FWWMR, OD, complete with pins and poles, Federal Stock No. 8340-257-2545.

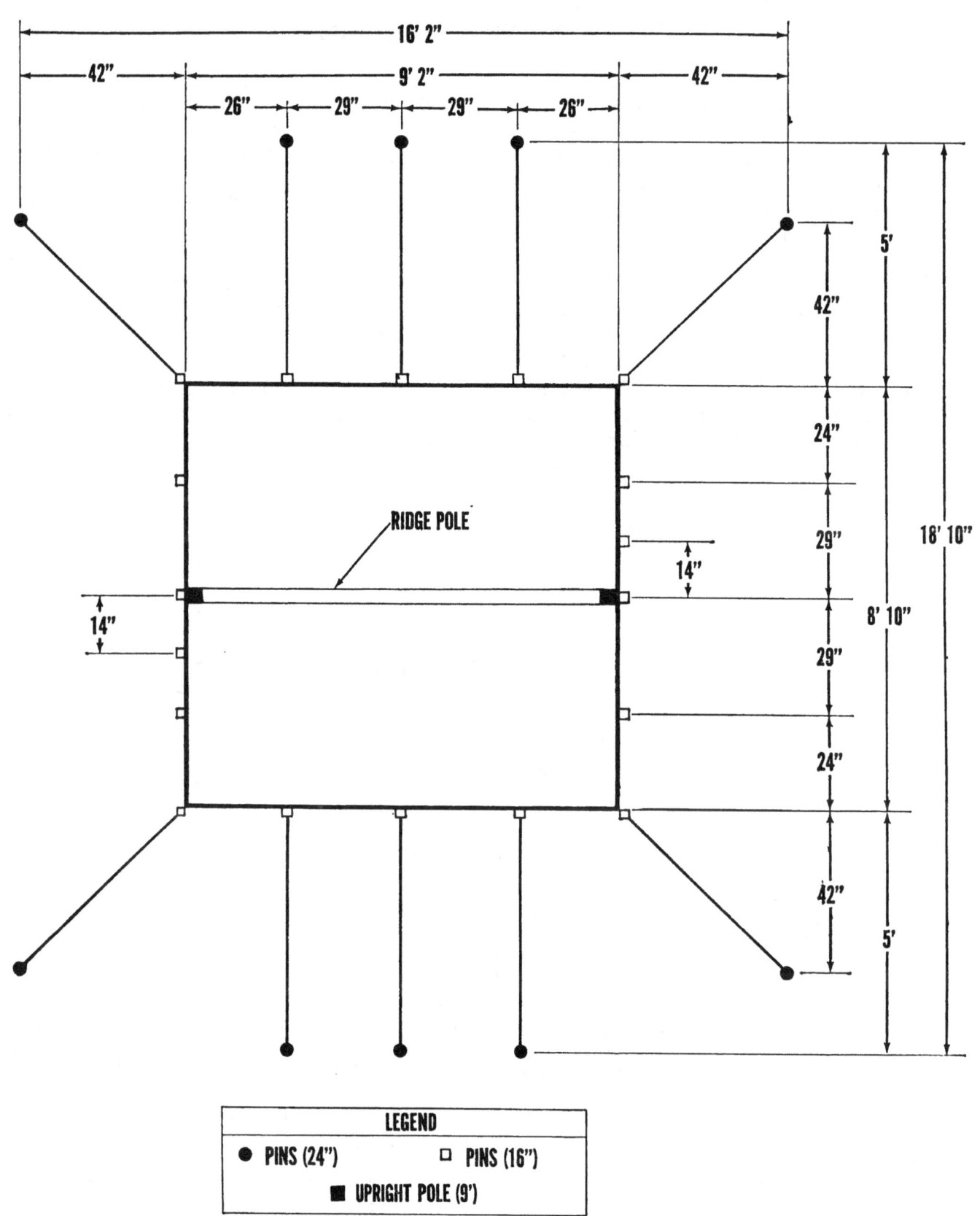

Figure 67. Ground plan of tent, wall, small, FWWMR, OD.

f. Ventilation. The tent is ventilated by 2 ventilator openings, one at each end section. The tent may also be ventilated by rolling up the side walls or by using the slits which form doors at each end of the tent.

g. Heating. The tent is heated by one M-1941 tent stove. Either of the ventilator openings may be used as a stovepipe opening. This permits the stove to be located at either end of the tent.

h. Cover. The tent is provided with a cover for use when in storage or when being transported.

i. Fly. A fly (Federal Stock No. 8340-188-9031) is available as a separate item of issue. The fly may be suspended above the deck of the tent to lower the temperature within the tent. It may also be pitched independently of the tent to provide quick shade and shelter. The fly measures 15 feet 6 inches by 9 feet 4 inches. It weighs 23 pounds and has a cubage in storage of 0.7 cubic feet. Six tent lines (line, tent, manila, one end sewed, one end with eye, 1/4-inch diameter, 8-feet long, Federal Stock No. 8340-252-2269) accompany the fly; they are used as eave lines.

77. Ground Plan

Before pitching small wall tent, study ground plan carefully (fig. 67).

78. Pitching

The small wall tent and the small wall tent fly are pitched in a manner similar to the large wall tent and the large wall tent fly (par. 72). The small wall tent, however, has no 45-foot guy lines (cut, spliced in center) or 4-foot 9-inch upright eave poles.

79. Striking

The small wall tent is struck similarly to the large wall tent (par. 73).

80. Folding

The small wall tent is folded similarly to the large wall tent (par. 74).

CHAPTER 3

PINS, POLES, AND LINES

81. Pins

a. Types of Pins Used. The types of pins (fig. 68) used with tends described in this manual are the 16-inch, the 24-inch, and the 36-inch wood pins, the 9-inch aluminum pins, and the 12-inch steel pins. Ordinarily, the 16-inch wood pins are used for footstops and the 24-inch wood pins are used for ridge and eave guy lines. The 9-inch aluminum pins and the 12-inch steel pins are used under cold-weather conditions. The quantity of each type pin required for different tents is shown in table I.

b. Method of Driving Pins.
  (1) All pins except the 24-inch guy-line pins and the 16-inch latrine screen pins are driven vertically into the ground. The 24-inch guy-line pins and the 16-inch latrine screen pins are driven into the ground at a 60° angle, with the top of the pin leaning toward the tent.
  (2) Wood pins are driven with the notches away from the tent.
  (3) Aluminum and steel pins are driven with the rounded portion of the pin away from the tent.

82. Poles

a. Poles (fig. 69) are of two types: upright and ridge. A ridge pole is usually fastened to two upright center poles by placing the spindles of the upright poles through holes at the ends of the ridge pole.

b. Poles are made of wood except the magnesium adjustable telescopic pole used in the 10-man arctic tent and the 5-man lightweight hexagonal tent.

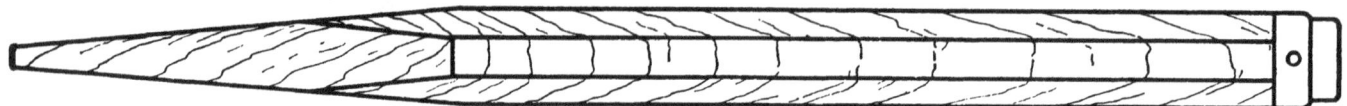
PIN, TENT, WOOD, 36-INCH

PIN, TENT, WOOD, 24-INCH

PIN, TENT, WOOD, 16-INCH

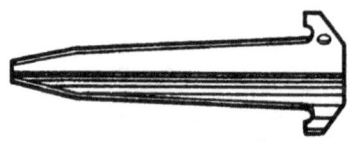
PIN, TENT, ALUMINUM, 9-INCH

PIN, TENT, STEEL, 12-INCH

Figure 68. Tent pins.

Table I. Pins Used for Tents

Type of tent	Pins*				
	16" wood (for footstops)	24" wood (for guy lines)	36" wood	9" aluminum (cold climate)	12" steel (cold climate)
Arctic, 10-man				28	
Assembly			39		
Command post	20	12			
General purpose, large	68	32		68	32
General purpose, medium	48	28		48	28
Hexagonal, 5-man				20	
Hospital, sectional	72	46			
Kitchen	32	31			
Latrine screen	8**				
Maintenance shelter	38	18			
Mountain				6	
Wall, large	26	18			
Wall, small	18	10			

*For Federal stock numbers of individual items, see sections on the various tents listed in manual.
**Used for guy lines; no footstops.

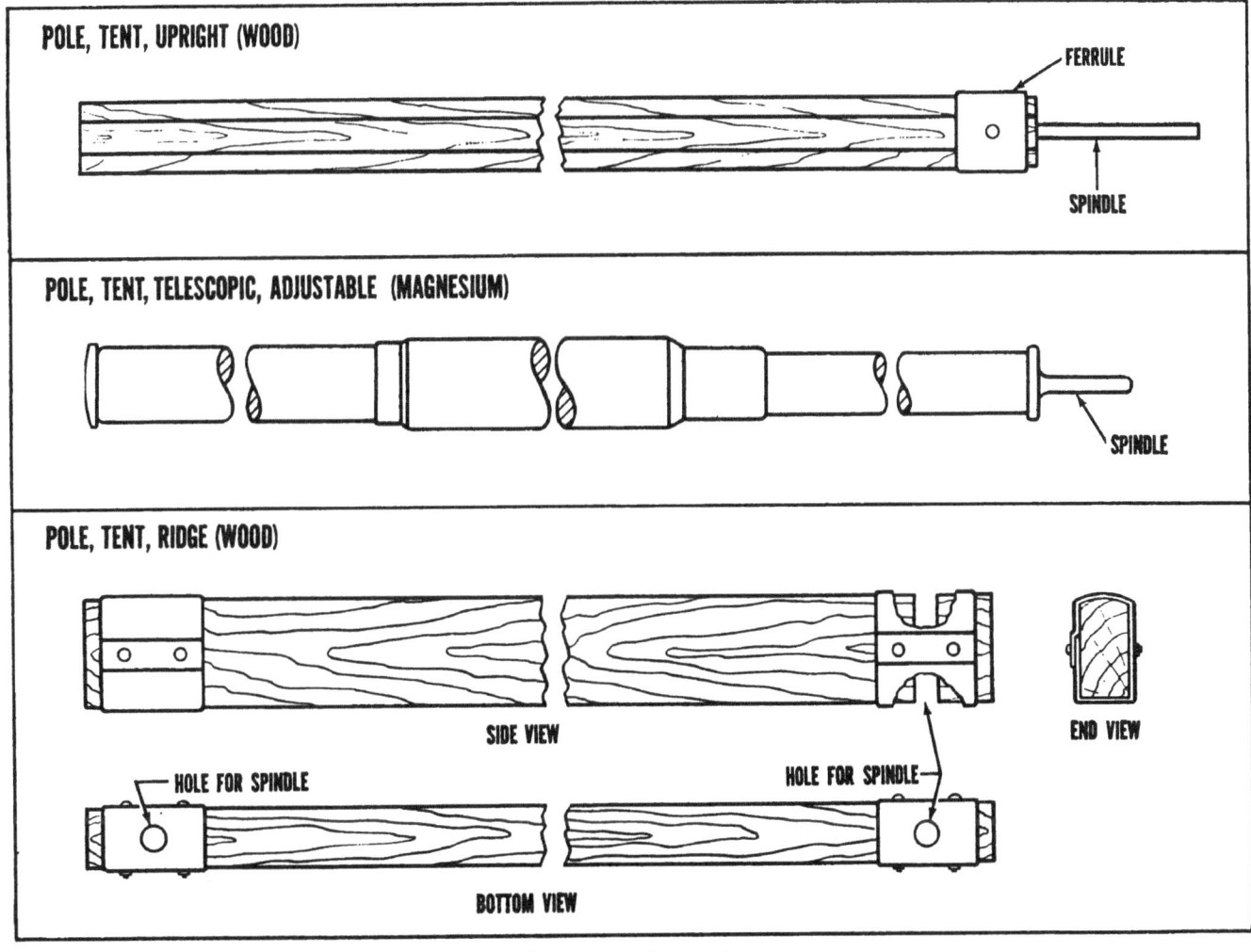

Figure 69. Tent poles.

c. Poles may be made of one piece or they may be made in sections which can be jointed.

d. Each pole or pole section is marked to show type, length, and section component; for example, "Upright—male section for 12 ft. 3 in. pole." This marking is important and should be taken into consideration in all cases in order to make sure that each tentpole is in its proper place.

e. When tents are being pitched, the upright poles are usually sunk from 2 to 4 inches into the ground.

f. The quantity of each type pole required for different tents is shown in table II.

83. Lines

Tent lines come in various sizes for different purposes on different tents. For best results, use the line prescribed for the purpose and the tent.

a. *Length of Lines.* Tent lines vary in length from a 19-inch-long footstop line to a 64-foot-long guy line.

b. *Diameter of Lines.* Tent lines vary in diameter from the 1/8-inch-in-diameter line used for sock lines and liner hoisting lines to the 1/2-inch-in-diameter line used as guy lines in the assembly tent.

c. *Tent Line Chart.* The quantity of each type

Table II. Poles Used for Tents

Type of tent	Poles*		
	Quantity	Type	Use
Arctic, 10-man	1	Telescopic, upright, jointed, magnesium, 4' 9" to 9'.	Center pole.
Assembly	3	Upright, solid, 21'	Center pole.
	30	Upright, solid, 8' 3"	Eave pole.
Command post	2	Upright, solid, 9'	Center pole.
	8	Upright, solid, 5' 8"	Eave pole.
General purpose, large	4	Upright, jointed, metal sleeve, 12' 3"	Center pole.
	4	Upright, solid, 6' 2"	Door pole.
	12	Upright, solid, 5' 8"	Eave pole.
General purpose, medium	2	Upright, jointed, metal sleeve, 10' 3"	Center pole.
	1	Ridge, jointed, metal sleeve, 17'	Ridge pole.
	4	Upright, solid, 6' 2"	Door pole.
	10	Upright, solid, 5' 8"	Eave pole.
Hexagonal, 5-man	1	Telescopic, upright, jointed, magnesium, 4' 9" to 9'.	Center pole.
Hospital, sectional	4	Upright, jointed, metal sleeve, 12' 3"	Center pole.
	22	Upright, solid, 6' 2"	Eave pole.
	4	Upright, solid, 8' 3"	Door pole.
Kitchen	2	Upright, jointed, metal sleeve, 12' 3"	Center pole (stack section).
	1	Ridge, solid, 5' 11¼"	Ridge pole (stack section).
	1	Upright, solid, 9'	Center pole (service section).
	1	Ridge, jointed, metal sleeve, 11' 10"	Ridge pole (service section).
	4	Upright, solid, 9'	Eave pole (stack section).
	6	Upright, solid, 6' 2"	Eave pole (service section).
	10	Upright, solid, 6' 2"	Awning pole (side).
	1	Upright, solid, 7'	Awning pole (front).
Latrine screen	6	Upright, solid, 7'	Side pole.
	1	Upright, solid, 7'	Entrance pole (inside).
	2	Ridge, solid, 9'	Ridge pole (long).
	1	Ridge, solid, 7'	Ridge pole (short).
Mountain	12	Pole section, tent, upright, male and female, 15" long.	End pole.
Wall, large	2	Upright, jointed, metal sleeve, 12' 3"	Center pole.
	1	Ridge, jointed, metal sleeve, 14' 2"	Ridge pole.
	4	Upright, solid, 4' 9"	Eave pole.
Wall, small	2	Upright, solid, 9'	Center pole.
	1	Ridge, solid, 9'	Ridge pole.

*For Federal stock numbers of individual items, see sections on various tents listed in manual.

Table III. Lines Used With Tents

Type of tent	Footstop, ¼" dia.	2', sewed 1 end, ¼" dia.	3' 4", sewed 1 end, ¼" dia.	3' 4", sewed 2 ends, ¼" dia.	4', unfinished 2 ends, ⅛" dia.	4' 6", sewed 1 end, ¼" dia.	4' 7", sewed 1 end, ⅜" dia.	6' eave line (mountain tent)	6', sewed 1 end, ¼" dia.	6', sewed 1 end, ½" dia.	8', sewed 1 end, ¼" dia.	8', w/eye, ¼" dia.	9', unfinished 2 ends, ⅛" dia.	10' 6", sewed 1 end, ¼" dia.	11', sewed 1 end, 5⁄16" dia.	11' 6", w/eye, 5⁄16" dia.	12', sewed 2 ends, ½" dia.	12' 6", unfinished, 7⁄32" dia.	12' 6", unfinished, yellow, 7⁄32" dia.	13' w/eye, 5⁄16" dia.	14', unfinished 2 ends, ⅛" dia.	14', sewed 1 end, ¼" dia.	15', guy line (mountain tent)	20', unfinished, 7⁄32" dia.	15', w/eye, ⅜" dia.	18' 9", unfinished, ⅛" dia.	18' 6", sewed 1 end, ½" dia.	19', unfinished, ⅛" dia.	19', sewed 1 end, ¼" dia.	19', w/eye, 5⁄16" dia.	21' 6", unfinished, 7⁄32" dia.	27', sewed 1 end, ¼" dia.	28' 6", unfinished, ⅛" dia.	30', unfinished, ⅛" dia.	30', unfinished 2 ends, ⅛" dia.	35', unfinished, ⅛" dia.	40' 6", unfinished, ⅛" dia.	45', sewed 2 ends, cut, spliced in center, 5⁄16" dia.	50', sewed 2 ends, cut, spliced in center, 5⁄16" dia.	52' w/thimble and hook, ½" dia.	64', sewed 2 ends, cut, spliced in center
Arctic, 10-man	14																	4	6					6		2					2		1				1				
Assembly			36				12		34	4							2			12							30					6								9	
Command post	24	2	10						2				2			8				6																					
General purpose, large	72		16						4											28		4																	2		
General purpose, medium	52		14						2											26		4																	2		
Hexagonal, 5-man	8																	5	6									1			1			1		1					
Hospital, sectional	130	28	8	16	30				22					4		30				8	8	4								8				6							
Kitchen	67			1					10				18	4		23				2	4									8											
Latrine screen			15									2			1										7																
Maintenance shelter	38		4	20		4					4					14				2		3							4			2									2
Mountain								4															2																		
Wall, large	26		4	24					4							14				2																		2			
Wall, large (fly)																6																									
Wall, small	18		2	20		4						12																													
Wall, small (fly)																6																									

*For Federal stock numbers of individual items, see sections on various tents listed in manual.

line required for different tents is shown in table III.

d. Knots. Four knots commonly used in tent pitching are the clove hitch, the round turn and 2 half hitches, the square knot, and the rolling hitch (fig. 70).

   (1) *Clove hitch.* The clove hitch is used to fasten a line to an anchorage. It will tighten as tension is applied, no matter which end of the hitch is pulled.

   (2) *Round turn and two half hitches.* The round turn and 2 half hitches is used to fasten a line to an anchorage. For permanency, the running end should be seized to the standing part.

   (3) *Square knot.* The square knot is used to join two lines of equal sizes.

   (4) *Rolling hitch.* The rolling hitch is used to fasten one line to another, especially a small line to a larger one.

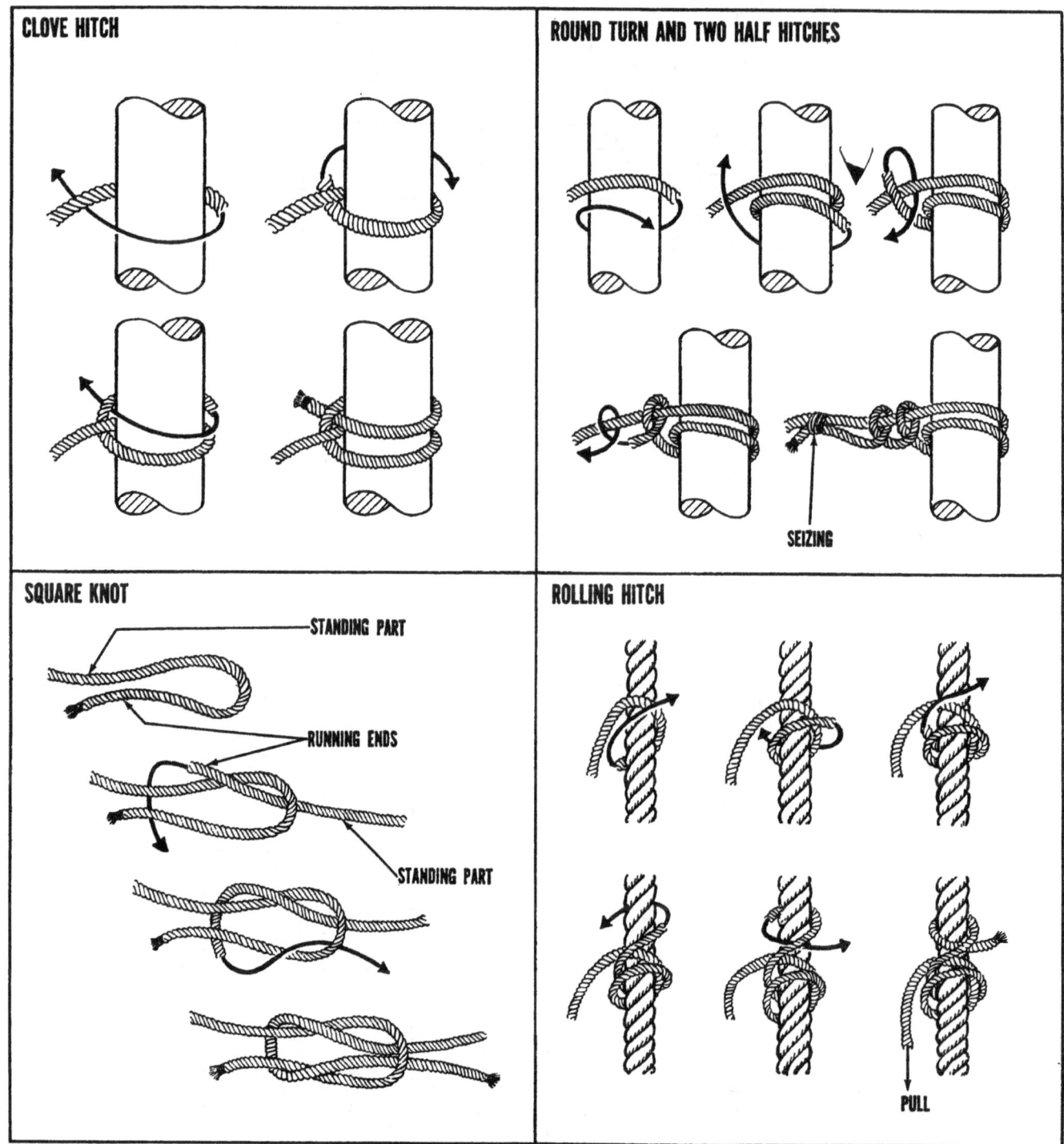

Figure 70. Knots used in tent pitching.

CHAPTER 4
SITE SELECTION

84. Choosing Tent Site

The following points should be considered in choosing a tent site:

a. The ground should be level and free from projecting tree roots and rocks. When such a spot is not available, a place can often be leveled and cleared with very little work. In the woods, moss and rocks may be used to level the ground.

b. The ground should be high enough for drainage. Drainage can be improved by trenching around tents and digging an outlet ditch to divert water in the desired direction (par. 86).

c. The tent should be protected from wind and storm.

d. An area having tough grass turf is desirable.

e. In the woods, the location should be away from dead trees or trees with large dead branches.

f. In hot weather, a shady area free from underbrush is desirable.

g. The tent should be placed far enough from a river, lake, or other body of water to be above the high-water mark.

h. In mountainous country, the tent should never be placed in a canyon or next to a dry creek bed. Such places have been known to fill up with rushing torrents in a remarkably short time. The tent should never be placed at the base of a cliff or steep mountainside, where there may be danger from avalanches and falling rocks.

85. Pitching Tent in Snow

a. Before selecting a campsite on snow-covered ground, prod surface with an ice or ski pole to see whether snow conceals any crevices. It may be impossible to find an area entirely without crevices, but it is possible to avoid accidents by knowing where they are.

b. When an adequate site on snow has been found, pack snow hard by stamping on it with skis or snowshoes, or better still, shovel top snow off until firm snow is found below.

c. Pitch tent so that entrance is not directly downwind. If the tent is pitched on snow with the entrance directly downwind, the entrance may become blocked, since snow tends to pile up in the lee of any object.

d. If site is not temporary, dig tent into snow. This will provide better protection from the wind. In open terrain with a strong wind, it may be necessary to build a snow wall on the windward side of the tent to protect it from the wind, thus making it easier to heat and less likely to blow down. Leave some space between sides of tent and snow wall in order to have room to shovel out snow that may drift into the tent.

e. When a tent is pitched on a slope, a horizontal platform should be formed. The snow which is removed may be packed around the outer edge of the platform to widen the space for the tent.

f. High winds, common in cold weather regions, require that tents be anchored securely. Tent pins may not provide sufficient anchorage. Arctic tents have snow cloths sewed along the bottom edge of tent walls. When an arctic tent is set up, snow cloths should be flat on the ground outside the tent. Place snow, snow or ice blocks, stones, logs, or other heavy objects on the cloths to help anchor the tent.

g. Do not attempt to drive tent pins into hard, frozen ground if the force required is excessive. Instead, chop small holes into the ground, insert tent pins into holes, and fill holes with slush or water; in a short time the tent pins will be firmly anchored. When removing pins from frozen ground, always chop them out; never hammer them sideways to break them loose.

h. Snow carried into a tent will melt and wet sleeping bags and clothing. The following precautions should be taken to keep snow out of tents:

   (1) Each man must take care to brush all snow from his clothing and boots before entering a tent.

   (2) One man should enter the tent first and take the sleeping bags, packs, and other

articles from the other men after the items have been brushed off completely.

86. Trenching Tent

a. A safe rule to follow is to always trench a tent. When the tent is pitched on heavy soil, clay, or a flat rocky surface, a trench should always be dug. When the tent is set up in very sandy soil, which absorbs water as fast as it falls, or when it is located on a mound which slopes off in all directions, a trench may not be necessary.

b. Dig trench all around tent (figs 71 and 72). Cut straight down, just outside footstop pins; do not dig in a V-shape. Slope the side away from tent inward toward the straight line.

c. Throw dirt from trench away from tent; never throw it against tent, for it will quickly rot the canvas.

d. In most cases, do not dig trench more than 4 or 5 inches deep and in the shallowest places not over 3 inches. There should be enough slope in the trench so that the water will flow freely toward the outlet and not back up.

e. To carry the water off, dig an outlet ditch (fig. 72) at the lowest point of the area and connect it to the trench which has been dug around the tent.

f. When there is a possibility that water may flow in from higher ground, dig a ditch to divert the water before it can reach the tent

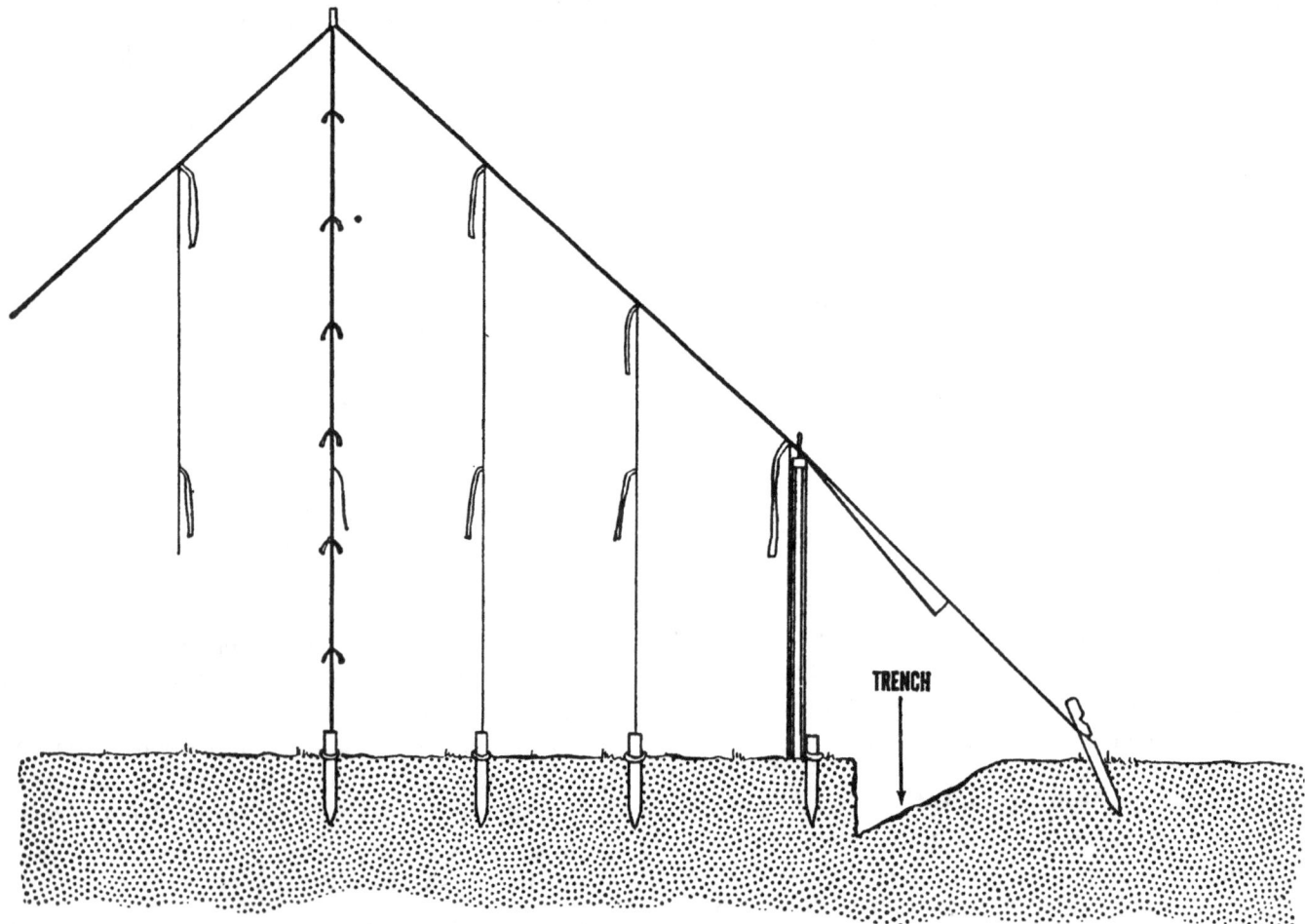

Figure 71. Cross-section view of tent trench.

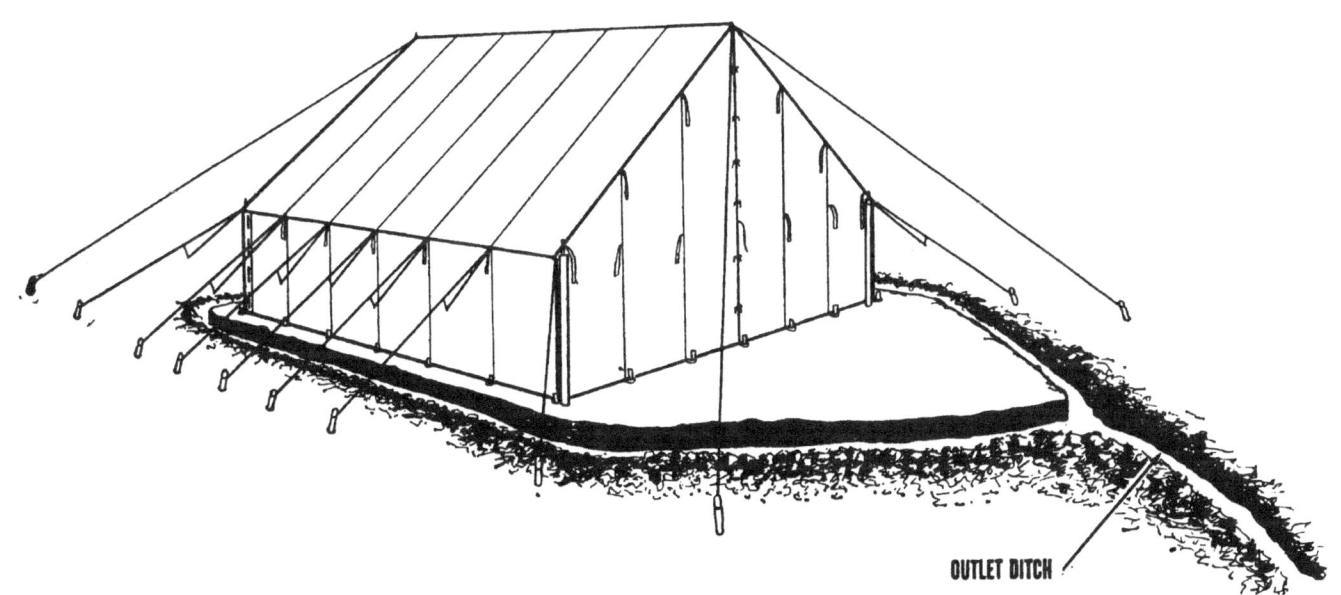

Figure 72. Trenching a tent.

CHAPTER 5

HEATING AND VENTILATION

87. Heating

a. Stoves and Heaters. The following stoves and heaters are available as separate items of issue to heat the tents described in this manual:

(1) *Stove, tent, M–1941.* The M–1941 tent stove may be operated with wood or coal, or with an oil burner attachment, using gasoline as a fuel. Conversion from one type of operation to another is quickly and easily accomplished. For information on the stove, see TM 10–725.

(2) *Stove, Yukon, M–1950.* The M–1950 Yukon stove uses leaded gasoline as fuel but may be adapted to burn coal or wood in an emergency for a short period of time. For information on the stove, see TM 10–735.

(3) *Heater, tent, gasoline, 250,000 B. t. u.* The 250,000 B. t. u. gasoline tent heater is an exterior heater operated with gasoline. For information on the heater, see TM 10–730.

b. Tent Heating Chart. The quantity of each type stove or heater for different tents is shown in table IV.

Table IV. Heating Equipment Used With Tents

Type of tent	Heating equipment		
	Stove, tent, M–1941	Stove, Yukon	Heater, tent, gasoline
Arctic, 10-man		1	
Assembly	*4		*2
Command post	1		
General purpose, large	3		
General purpose, medium	2		
Hexagonal, 5-man		1	
Hospital, sectional	**1		
Kitchen			
Latrine screen			
Maintenance shelter			1
Mountain			
Wall, large	1		
Wall, small	1		

*Two gasoline tent heaters or four M–1941 tent stoves may be used.
**One per middle section.

c. Stovepipes and Stovepipe Openings.

(1) A stovepipe of fixed length and a draft diverter are issued as component parts of the M–1950 Yukon stove. Six 2-foot lengths of stovepipe are issued as component parts of M–1941 tent stove. A spark arrester is available and must be used whenever solid fuel is used in an M–1941 tent stove. An oil-burner kit, Burner, oil, stove, tent, M–1941, is available to convert an M–1941 tent stove to an oil-burning heater. A draft diverter is issued as a component part of the oil-burner kit. No M–1941 heater should be operated without either a draft diverter or a spark arrester, dependent upon the fuel used. The diverter and arrester should NEVER be used at the same time. Additional 4-inch stovepipe and 4-inch adjustable elbows are available as separate items of issue.

(2) Stovepipe openings are built into most tents; some openings are reinforced and the tent protected against the head of the stack, while others are not protected. Metal shields are available, which should be placed in the stovepipe openings of tents where there is no reinforcement or heat protection for tent material. Stovepipe openings have canvas flaps attached which may be closed for protection against the weather and left open for ventilation when stoves are not in operation.

d. Heating Individual Shelters. Normally, there is no provision for heating the 2-man mountain tent. However, when men are forced to stay in it for long periods of time or when the men are wet and need to dry off, one or more of the following expedient measures may be used:

(1) A brush fire may be built over the area on which the tent is to be pitched and kept going for an hour or two. Then, the area should be cleared of all coals and sparks and the tent set up. The

ground will remain warm for several hours and the earth will be dry to sleep on.

(2) Stones 5 or 6 inches in diameter may be put in a hot fire for 2 or 3 hours, then rolled or lifted with forked sticks into the tent. If a bucket or other metal container is handy, it may be used to hold the rocks or it may be placed upside down over them. The rocks will continue to give off heat for several hours. If there is not sufficient room to pile hot rocks in the tent safely, dig a hole and fill it with hot rocks even with or slightly below the surface of the ground.

(3) Although the one-burner cooking stoves issued are intended for cooking purposes, their heat will also take the chill off the inside of individual shelters. However, be extremely careful of this method of heating the 2-man mountain tent because of the danger of carbon monoxide.

(4) A gasoline lantern is an excellent heater, and even candle lanterns will take off the chill.

88. Ventilation

It is extremely important that a tent be ventilated properly. All tents have ventilating facilities provided. Most have built-in ventilators of various types. When stoves are not being used, stovepipe openings may be used for additional ventilation. In hot weather, the doors may be opened, and on most tents the side walls may be rolled up, or the windows opened, to give a free circulation of air. The air coming in around the bottom of the tent should never be depended upon for ventilation. If the sod cloth or snow cloth is properly weighed down, very little air will enter. The bottom edge of the tent is the least desirable from which to get ventilation. It is like trying to ventilate a house through the cracks in the floor.

CHAPTER 6
GROUND COVERINGS

89. Sod Cloths

Sod cloths are strips of canvas sewed along the bottom edge of tent walls. When a tent is set up, sod cloths should be flat on the ground inside the tent and should be weighted down with clean dry material. The purpose of sod cloths is to keep out wind and cold, as well as insects and small animals. Banking a tent inside with leaves or earth is a poor substitute for sod cloths. The leaves and earth will not stay tight against the tent, and damp earth will rot the canvas.

90. Ground Cloths

a. Ground cloths are not generally issued for use with tents. Because earth is soon trodden down hard so that it is easy to sweep and keep clean, it is more sanitary to use as a floor than ground cloths, which get dirty and are hard to keep clean.

b. The 2-man mountain tent has a built-in cloth floor. This is desirable because the tent is used in extremely cold climates and the built-in cloth floor helps keep out cold wind and drifting snow.

c. The maintenance shelter tent comes equipped with 6 ground cloths which may be used to form a floor. Tanks and trucks are repaired in this tent and it is advisable to have the ground cloths as a floor so that the spare parts and tools can be laid down on them rather than on the ground.

91. Snow Cloths

See paragraph 85*f*.

92. Wooden Floors

Wooden floors are used most frequently with general-purpose and wall tents pitched at semi-permanent installations. It is very important that the floor should follow the ground plan of the tent. If it is too large, it will pull the tent out of shape. Even a water-resistant tent will shrink slightly after a hard rain and as a result will be a few square inches smaller than its specified dimensions. Tent floors should be built in sections to facilitate transportability from one location to another.

CHAPTER 7

CARE OF TENTAGE

Section I. PROTECTION OF TENTS AGAINST DAMAGE

93. General

Probably the greatest amount of damage to tentage is caused by carelessness, such as forgetting to loosen the lines when it starts to rain, not bothering to use spark arresters or draft diverters, adjusting lines carelessly, driving pins in a slipshod manner, or dragging tents over rough ground. To prolong the life and usefulness of tentage, observe the following rules:

a. Pitch, strike, and fold tentage in the manner described in this manual. Do not try to take shortcuts unless sure that no damage will be done. In order to protect the top of the tent during handling and in storage, fold so that the side walls rather than the top of the tent will be exposed.

b. Observe utmost care when pitching and striking tents, making sure the material does not tear on protruding pins, overhanging branches, or other objects.

c. Never drag a tent along the ground or floor.

d. Be sure to use all the necessary parts and accessories for each tent and to use them for their intended purpose.

e. Pack tents carefully for shipment. Some tents are issued complete with bag or cover. In this case, be sure to carry tent in bag or cover. When no bag or cover is issued, the tent is usually received wrapped in osnaburg or burlap. Save this material for rewrapping when the tent has to be moved again. Normally, a tent should never be transported without a covering of some kind.

f. Pack pins and poles separately from tent itself except in the case of arctic tents and individual shelters.

g. Inspect tentage at frequent intervals to make sure that it is in serviceable condition. Particular attention should be given to seams, bindings, lines, and all places where strain is exerted. Be constantly on the lookout for—

(1) Any evidence of mildew.

(2) Any foreign matter which may have collected on the tent.

(3) Small rips and holes, splitting of seams, grommets which have become loose, lines which are beginning to rot, or anything else which does not appear to be in normal condition.

94. Protection Against Rain

a. Most tents are water repellent. However rain causes tent canvas and lines to shrink, the shrinkage often becoming sufficient to tear the tent. Tents have been torn completely in two under such circumstances.

b. Before tent lines become water soaked, loosen them sufficiently so that when they shrink they will not become tight enough to tear the tent. To compensate for shrinkage, eave and corner lines should have a free swing of approximately 18 inches at the middle of the line.

95. Protection Against Wind

In a strong wind, tighten all lines immediately. Close door entrances, secure walls to footstop pins, and close all corners.

96. Protection Against Fire

a. Most tents are fire resistant. This does not mean that they will not burn; they usually do not burst into flame, but smolder and char.

b. When using a stove in a tent, every precaution must be taken to avoid fires. Spark arresters or draft diverters must be installed and shields placed around stovepipe openings. All personnel should be well trained in building and maintaining stove fires and should be familiar with all fire regulations.

c. Whenever possible, fire extinguishers containing water should be kept in the tent area.

97. Protection Against Mildew

a. Most tents are mildew resistant. This does not mean that they are not subject to mildew. Under warm and damp conditions, especially in tropical and jungle areas, tents may be ruined by mildew in a few days, if proper care is not taken.

b. To prevent mildew, follow these rules:

(1) Never fold or roll a tent when wet. Even if it is only damp from dew, it will mildew when stored. Make doubly sure that the seams and edges of the tent, especially the bottom edge and the sod cloth, are dry.

(2) When storing or transporting, keep pins and poles separate from tents, except those used with arctic tents and individual shelters, which are cleaned and dried before being placed with the tent.

(3) Keep tents clean at all times. If a tent is pitched under trees, inspect the tent roof frequently to see if it is being harmed by drippings from branches or leaves. The growth of fungi and mold is caused to some extent by tree drippings, oils, greases, and starches, which accumulate on tentage.

(4) Before storing, dry a tent by hanging it up off the ground in bright sunlight. A tent dried on the ground or left hanging outdoors after sundown might absorb enough dampness for mildew to start. When necessary, a tent may be dried indoors. When drying indoors, hang the tent in a well-ventilated place, high enough to permit the tent to be suspended off the floor.

(5) Do not drag tentage along the ground or permit it to come in contact with the ground while in storage.

(6) When storing tents, stack them on dunnage supported by 2- by 4-inch lumber.

   (*a*) If the floor is hard surfaced or wooden, the tentage should be at least 4 inches from the floor.

   (*b*) If the floor is earthen, the tentage should be at least 8 inches from the ground.

   (*c*) Only lumber that has been thoroughly cured should be used for dunnage, since the moisture contained in green lumber will promote the growth of mildew.

   (*d*) When dampness in the atmosphere is prevalent, dunnage should be used between each course to permit circulation of air between the blocks. The blocks should be separated and reduced to a minimum number of courses to permit passage of air on all four sides.

(7) When tents are to be stacked near ventilators or openings that may admit moisture, protect tents by packing them in bags or waterproof coverings.

(8) Do not place tentage received from the field in bags until tents are thoroughly dried and all dirt removed by stiff brushes. If any visible signs of mildew are present, hang tents in open air, preferably in the sun.

(9) Give priority of issue to tentage that has been in storage the longest. To prevent issue of newly stored tentage before older stocks are exhausted, blocks should be marked in accordance with length of time tentage has been in storage.

(10) When tentage is stored in open sheds or in tents, it should be staked well away from sides and ends of shelter (preferably about 20 feet), and items not affected by moisture should be stacked between tentage and outer edges of shelter.

(11) Withdraw from storage tentage found to be infected with mildew. Brush with a stiff brush, allow to dry thoroughly, and issue immediately to installations where driest atmospheric conditions prevail. If there is no opportunity for immediate issue, segregate infected tentage from sound tentage to prevent contamination. Tents which have become unserviceable should be turned in to a salvage installation for classification, repair and return to stock, or for destruction.

Section II. PROTECTION OF PINS, POLES, AND LINES AGAINST DAMAGE

98. Pins

All wooden tent pins currently issued receive a wood-preservative treatment. Care should be taken in handling pins to see that they are not broken or otherwise damaged. In determining the serviceability of pins, look for cracks, splits, distorted ends, and broken or flattened points.

99. Poles

Care should be taken in handling tentpoles to see that they are not broken or otherwise damaged. In determining the serviceability of poles, look for cracks, splits, condition of metal joiners, and missing or bent spindles.

100. Lines

Lines should be inspected frequently. The stability and safety of the tent may depend on the condition of the various lines used—guy lines, eave lines, footstops, door fasteners, and others. Deterioration in tent lines is of two kinds: physical and chemical. Physical damage is caused by surface wear or from internal friction between the fibers. Chemical damage is caused by exposure to weather conditions and acids. To prevent damage to tent lines, observe the following rules:

a. Store lines properly in a dry, unheated building or in a room with free air circulation. Place lines in loose coils off the floor on wooden grating, or hang them on wooden pegs. It is best to hang small lines in loose coils and to coil large sizes loosely on a grating or platform raised from the floor to insure necessary circulation of air. Never store lines in a small confined space without air circulation. Clean thoroughly before storing. Continuous exposure to sunlight is injurious to lines. Improper storage conditions frequently cause dry rot.

b. Dry lines properly after exposure to dampness. Lines are best dried when hung loosely between two trees or other objects so that they do not come in contact with the ground.

c. Keep lines clean. If lines become dirty, they should be washed in clean water and thoroughly dried. Grit from sand, mud, or other materials, if allowed to remain and work into lines, will grind and wear the fibers.

d. Protect lines from chemicals. Keep lines away from chemicals or their fumes, especially acids or alkalis. Drying oils, such as linseed oil, and paint will also damage lines.

e. Slack off guy lines. When guy lines or other supports are exposed to the weather, slack them off to prevent overstrain because of shortening from wetting.

f. Reverse lines, end for end, periodically, so that all sections of the lines will receive equal wear. When wear is localized in a short section, periodical shortenings will present a new wearing surface.

g. If a line becomes damaged, cut and splice. A good splice is safer than a damaged section.

h. Whip ends of lines to prevent raveling.

Section III. REPAIR METHODS

101. Cement Patches

Cement patches are used for minor repairs in the field.

a. Materials Used.
   (1) Cement.
   (2) Circular patches.
   (3) Roller.
   (4) Brushes.

b. Size of Patches. Three sizes of circular patches are used to repair tents.

Patches		Diameter or length of hole or rip
Size No.	Diameter	
	Inches	Inches
1	3	1½
2	4⅜	2⅞
3	6¼	4¾

c. Procedure.
   (1) In selecting the size of patch to use, allow

a minimum overlap of three-fourths of an inch on all sides of the hole or rip.
(2) Remove foreign materials, such as dirt and grease, completely from the part of the tent to be repaired.
(3) Apply a thin coat of cement to the portion of the tent to be covered by the patch and to the entire surface of the patch. Allow the first coat to dry to a tacky state (approx. time: 5 min.), and then apply a second coat of cement. While surfaces are still wet, press both surfaces together by means of the roller provided for this purpose.

102. Hand Sewing

Hand sewing is done mostly in the field when sewing machines, grommet-setting dies, and other essential equipment are not available. Hand-sewed patches, when properly stitched, will outlast a patch put on by machine. A patch must be stitched on any hole exceeding 4¾ inches in diameter.
  a. Materials used.
  (1) No. 15 sailmaker's needle.
  (2) Five-ply cotton twine.
  (3) Beeswax.
  (4) Sewing palm.
  (5) Olive-drab cotton, duck, 12.29-ounce-per-square yard, fire-, water-, weather-, and mildew-resistant-treated.

  b. Procedure. Cut a piece of duck large enough to overlap the hole (at least 1 in. on each edge plus one-half inch for turning under the ragged edge). Prepare patch by turning under the four ragged edges approximately one-half inch and sewing them to hold turned edges in place. With sewing palm and a needle with two strands of well-waxed cord, and using the sailmaker's stitch, sew patch onto tent, beginning at top right-hand corner. For the first stitch, push needle down through edge of patch (and through tent below) and out through patch at a point one-fourth inch diagonally below point of entry. For the second stitch, push needle through top of patch again, at a point one-half inch from original point of entry. Continue in the same manner until patch has been sewed tightly against tent. A patch sewed with this stitch is water repellent.

103. Machine Sewing

When the hole or rip is so large that machine sewing is necessary, send tent to nearest repair shop without delay.

APPENDIX I
REFERENCES

1. Publication Indexes

DA PAM 108-1	Index of Army Motion Pictures, Television Recordings, and Filmstrips
DA PAM 310-1	Index of Administrative Publications
DA PAM 310-2	Index of Blank Forms
DA PAM 310-3	Index of Training Publications
DA PAM 310-4	Index of Technical Manuals, Technical Regulations, Technical Bulletins, Supply Bulletins, Lubrication Orders, and Modification Work Orders
DA PAM 310-7	Index of Tables of Organization and Equipment, Reduction Tables, Tables of Organization, Type Tables of Distribution, and Tables of Allowances

2. Publications Relative to Tentage

FM 21-15	Individual Clothing and Equipment
TM 10-250	Storage of Quartermaster Supplies.
TM 10-269	Repair of Canvas and Webbing
TM 10-616	Shelter, Tent-Type, Portable, Sectional
TM 10-633	Canvas Repair Kit
TM 10-725	Stove, Tent, M1941, Complete, and Burner, Oil, Stove, Tent, M1941
TM 10-730	Heaters, Tent, Gasoline, 250,000 B. t. u.
	Herman-Nelson (Model GT-3077) and Silent Glow
TM 10-735	Stove, Yukon, M1950
QM 5-24	Supply Manual
FB 150	Care and Preservation of Tentage
FS 7-19	Pitching and Striking the Wall Tent
FS 7-22	Pitching and Striking the Latrine Screen
FS 8-105	Pitching and Striking the Sectional Hospital Tent

APPENDIX II
GLOSSARY

Bail ring or chains and supporting rings—That part of an assembly tent by means of which the lifting block and tackle is rigged to the main center poles and to which the various sections of canvas are lashed.

Bar, double-eye—A metal bar with two loop eyes, one for the attachment of a guy line and the other for the tentpole spindle. It is used with stress distribution patches (see below) for the transmission of stresses in a radial pattern instead of in a concentrated one.

Becket—A loop, attached in a series along the edge of a tent section, used for chain-lacing two sections together.

Bull's-eye, strapped—A device consisting of a wire loop encircling a wood ring, through which hoisting lines for raising tent liners are reefed.

Chape—Any loop of leather or metal.

Corner strap (webbing)—A strap which secures the D-ring or triangle to the corner of the tent.

Cut splice—A loop spliced in the center of ridge guy lines to fit over the spindle of the pole which supports the top of the tent.

Door flap—That part of a tent which forms the covering to the entrance.

Eave line—The supporting rope line that extends from the eave of a tent to a pin driven into the ground.

End wall—Either end of a tent extending from the ridge or eave to ground line.

Extension cloth—A canvas strip added to one end of the roof of a tent to prevent leakage between units when two or more tents, or sections of the same tent, are put together as one. The extension cloth overlaps the tent or section of tent next to it.

Fair-lead—A metal device to which eave guy lines are attached. The fair-lead transfers the guy line pull to the triangle square, or diamond with hook, which in turn transfers the pull to the webbing framework of the tent.

Ferrule—A ring or cap, made of metal, put around a tentpole to strengthen it or to prevent splitting and wearing. It is generally used on pole tops where the spindle is inserted. It is also used to reinforce the tops of 36-inch wood pins.

Footstop—A rope loop inserted through a grommet at the bottom of a side wall. This loop, fitted over a pin driven into the ground, secures the side wall.

Grommet—A metal eyelet consisting of two parts, grommet barrel and washer, clamped together securely through each side of the material, forming a ring. Grommets are used for attaching tie lines, eave lines, and supporting lines. They can be of two types: sheet grommet, the barrel of which can be used either with a plain, flat sheet washer or a toothed washer; and spur grommet, made, according to size, of a heavy-gage metal with a rolled rim, with teeth attached.

Jumper line—A line, one end of which is permanently fastened to the deck of the tent and the other tied to an upright pole, to prevent the canvas from jumping off the spindle of the pole, especially during stormy weather.

Lacing line (corner)—The line used to lace together the side walls of a tent at the corners.

Lanyard—Any length of rope, regardless of size, hand-whipped on one end and reinforced by means of spliced eye and thimble. It is used to make various lashings.

Line—A length of rope composed of a certain number of strands, each strand made up of a number of single threads.

Liner—An inner liner, slightly smaller than the tent itself, used to create a dead-air space between the tent and liner and serving as an insulating medium.

Lug—A piece of canvas which is doubled over with the edges folded under and which is stitched to the tent to hold ropes or lines. It is often sewed next to the doorway of a tent to hold the line used to secure the door flap when it is open.

Pin (stake)—Wooden or metal peg driven into the ground. Tent lines are attached to pins to aid in holding the tent in position.

Plate—Thin piece of metal, through which the spindle of the tentpole projects and to which sections of the supporting webbing framework of the tent are attached, either directly or by means of triangle with hook. It is used at the ridge and eave of a tent. By allowing fullness in the canvas around the plate, the tensile stresses are imposed upon the webbing instead of on the canvas.

Reinforcement—Patch of canvas used to strengthen a seam or a point of abrasion and tension.

Sash (window)—A flexible vinyl film window used with tentage; the window is detachable and is installed in the tent by means of grommets and snap fasteners.

Side wall—That part of the tent extending from below the eave to the ground on all sides of the tent. Depending on the design of the tent, it may be either part of or separate from the top.

Slip—A device used to adjust the eave or ridge guy lines on tents. Usually slips are 4 inches long, made of steel wire $3/16$ inch in diameter, one end twisted like a coil spring and the other end looped to form an eye. Slips for lightweight arctic tentage are made of 4-inch magnesium bars, $1/4$-inch thick and $7/8$-inch wide, with a hole at each end for the entrance of the tent line.

Sock line—A line provided in lightweight arctic tentage upon which light articles of clothing may be hung and dried.

Stress distribution patch—A triangular-shaped canvas patch, containing a double-eye bar, stitched on the eave of wall tent flies to distribute tensile strength through the canvas in a radial pattern (instead of in a concentrated one, such as exists around a grommet or handworked ring).

Tent top or tent deck—Any part of the tent above the side wall or eave.

Thimble—A metal insert, generally egg-shaped and ranging in size from $1\frac{1}{4}$ to 4 inches, which fits into a spliced eye. Thimbles are used to reinforce the ends of ropes, lanyards, and tie lines.

Triangle with hook, square with hook, or diamond with hook—A metal device used with a fairlead or plate to transfer the stress on the tent canvas or on a guy line to the webbing framework of a tent.

Wall line—A piece of webbing sewed into the seam of a tent, with half of the line inside and half outside the tent. This line is used to tie up the side wall when it is raised.

INDEX

	Paragraphs	Page
Arctic tent, 10-man	3–8	4
Assembly tent	9–14	9
Basis for tent issue	2	3
Command post tent (M-1945)	15–20	17
Damage, protection against	93–100	99
Fire, protection against	101	101
Floors, wooden	97	100
Fly, tent:		
Wall, large	70, 72	77, 79
Wall, small	76	85
General purpose tent, large	21–26	25
General purpose tent, medium	27–32	31
Glossary	App. II	104
Ground cloths	90	98
Hand sewing	102	102
Heaters	87	96
Hexagonal tent	33–38	41
Hospital tent, sectional	39–44	46
Kitchen tent, flyproof	45–50	54
Knots	83	90
Latrine screen	51–56	61
Lines:		
Protection against damage	100	101
Types	83	90
Machine sewing	103	102
Maintenance shelter tent	57–61	65
Mildew, protection against	97	100
Mountain tent, 2-man	62–66	72
Patches:		
Cement	101	101
Hand-sewed	102	102
Pins:		
Protection against damage	98	101
Types	81	88
Poles:		
Protection against damage	99	101
Types	82	88
Rain, protection against	94	99
Red Cross marker:		
Large	67	76
Small	68	76
References	App. I	103
Repair methods	101–103	101
Screen, latrine	45–50	54
Site selection	84–86	93
Snow, pitching tent in	85	93
Sod cloths	89	98
Stoves	87	96
Trenching tent	86	94
Ventilation	88	97
Wall tent:		
Large	69–74	77
Small	75–80	85
Wind, protection against	95	99
Wooden floors	92	98

[AG 424.1 (21 Oct 55)]

By Order of *Wilber M. Brucker*, Secretary of the Army:

MAXWELL D. TAYLOR,
General, United States Army,
Chief of Staff.

Official:
    JOHN A. KLEIN,
Major General, United States Army,
The Adjutant General.

DISTRIBUTION:
    Active Army:

Tec Svc, DA (5) except TQMG (25)	Div (5)	MFSS (300)
Tec Svc Bd (2)	Brig (2)	PMST (1)
Hq CONARC (5)	Regt/Gp (2)	Gen Depots (2)
Army AA Comd (2)	Bn (1)	QM Sec, Gen Depots (10)
OS Maj Comd (10)	Co (1)	QM Depots (10)
OS Base Comd (10)	Ft & Cp (2)	Trans Terminal Comd (2)
Log Comd (1)	USMA (15)	OS Sup Agencies (2)
MDW (2)	Gen & Br Svc Sch (20) except	Mil Dist (3)
Armies (15)	Southeastern Sig Sch (50)	QM R & D Comd (3)
Corps (5)	QM Sch (25)	Arty Cen (800)

NG: State AG (6); units—same as Active Army except allowance is one copy per unit.
USAR: None.
For explanation of abbreviations used, see SR 320-50-1.

www.ingramcontent.com/pod-product-compliance
Lightning Source LLC
Chambersburg PA
CBHW081232080526
44587CB00022B/3918